D0466718

TOXIC DECEPTION

The Center for Public Integrity

TOXIC
DECEPTION

**How the Chemical Industry
Manipulates Science,
Bends the Law, and
Endangers Your Health**

**Dan Fagin, Marianne Lavelle,
and the Center for Public Integrity**

A BIRCH LANE PRESS BOOK
Published by Carol Publishing Group

To Alison; for Anna and Lily—D.F.

For Moira, Claire, and all their cousins, present and future—M.L.

Copyright © 1996 Dan Fagin, Marianne Lavelle, and the Center for Public Integrity
All rights reserved. No part of this book may be reproduced in any form, except by a
newspaper or magazine reviewer who wishes to quote brief passages in connection with a
review.

A Birch Lane Press Book
Published by Carol Publishing Group
Birch Lane Press is a registered trademark of Carol Communications, Inc.

Editorial, sales and distribution, rights and permissions inquiries should be addressed to
Carol Publishing Group, 120 Enterprise Avenue, Secaucus, N.J. 07094.

In Canada: Canadian Manda Group, One Atlantic Avenue, Suite 105, Toronto, Ontario
M6K 3E7

Carol Publishing Group books may be purchased in bulk at special discounts for sales
promotion, fund-raising, or educational purposes. Special editions can be created to
specifications. For details, contact Special Sales Department, 120 Enterprise Avenue,
Secaucus, N.J. 07094.

Manufactured in the United States of America
10 9 8 7 6 5 4 3 2 1

Library of Congress Cataloging-in-Publication Data

Fagin, Dan.
    Toxic deception : how the chemical industry manipulates science,
  bends the law, and endangers your health / Dan Fagin, Marianne Lavelle, and the Center
  for Public Integrity.
        p. cm.
    "A Birch Lane Press book."
    Includes bibliographical references.
    ISBN 1-55972-385-8 (hc)
    1. Environmental toxicology—United States.   2. Chemical industry—United
  States. I. Fagin, Dan. II. Title.
  RA1226.L38   1996
  615.9'02'0973—dc20
                                                                            96-41045
                                                                                CIP

Contents

Foreword

On March 31, 1993, *Frontline*, a Public Broadcasting System television program, aired a provocative documentary, "In Our Children's Food," about the harmful effects of pesticides and the general inability of the federal government to regulate the widespread use of cancer-causing farm chemicals. The documentary, narrated by Bill Moyers and produced by Martin Koughan, later received an Emmy award from the National Academy of Television Arts and Sciences. It was also honored by Investigative Reporters and Editors as the best TV documentary in the United States in 1993.

The morning after the program aired in most U.S. cities, however, a remarkable editorial appeared in the *Wall Street Journal*. Titled "'Frontline' Perpetuates Pesticide Myths," it was written by Dennis T. Avery, who was identified as a "fellow at the Hudson Institute" and "director of Hudson's Center on Global Food Issues." The author, an agricultural economist whose work is funded by agribusiness interests, charged in his opening paragraph that the *Frontline* documentary had "made recommendations that would increase our cancer and heart disease rates, increase the risk of world hunger, and plow down millions of square miles of wildlife habitat." Avery wrote about the "ignorance" of Rachel Carson, the respected author of the classic 1962 book *Silent Spring*, and also lamented the government's banning of the pesticide DDT, which, he said, was not "dangerous to people or birds."

What most intrigued us was not the vitriol of Avery's editorial but its curious timing. No one places an op-ed article in a major newspaper the night before its publication; indeed, the *Frontline* program ended at 10 P.M., Eastern time. How did Avery do it?

vii

That question gnawed away at producer Koughan so much that he telephoned Avery directly. Koughan discovered to his dismay that the editorial actually had been submitted to the *Wall Street Journal* a full *week* before the *Frontline* program was broadcast. He later ascertained that the pesticide industry had somehow obtained a copy of "In Our Children's Food" a *month* before the air date. An entire public-relations campaign had been waged against the show; individual PBS stations had been contacted, and the American Cancer Society had been persuaded to send out a nationwide bulletin to its members, advising them to disregard the information in the program. As Sheila Kaplan reported in *Legal Times*, the American Cancer Society instructed its branch offices how to respond to public inquiries, asserting that "the program makes unfounded suggestions...that pesticide residues in food may be at hazardous levels."

How did all of this happen? Porter/Novelli, a public relations and lobbying firm based in New York City—representing major chemical companies such as Rhone-Poulenc Ag Company, DuPont, and Hoechst-Roussel, in addition to the pesticide industry–financed Center for Produce Quality—had quietly been doing pro bono PR work for the American Cancer Society. When word became known that a major, critical documentary was about to be telecast, the firm quickly sprang into damage-control action, orchestrating the American Cancer Society's "reaction" to the program. It was a public relations coup, one cited by critics of "In Our Children's Food" as evidence that the dangers to children from pesticides had been overstated.

A postscript to this episode is that weeks after the *Frontline* documentary, in response to an important, long-awaited report by the National Academy of Sciences about the health risks of pesticides to children, then–Secretary of Agriculture Mike Espy, Environmental Protection Agency Administrator Carol Browner, and Food and Drug Administration Commissioner David Kessler made an unusual joint announcement. In what Kessler called "a major landmark in the history of food safety," the three officials pledged that they would initiate legislative and regulatory reforms to reduce pesticide use in the United States.

Nearly four years after this dramatic, heavily reported announcement, however, the EPA has not ordered a single pesticide removed from the market (though some have been voluntarily withdrawn by their manufacturers for safety reasons).

Why is it that when it comes to regulating harmful chemicals in this nation, for years and years government officials—Democrats and Republicans alike—have seemed to be swimming in quicksand?

At the Center for Public Integrity, we are not scientists, and we do not advocate public policies. We are a not-for-profit, 501(c)(3) investigative research organization that examines public service and ethics-related issues, and we have published 20 investigative studies since we opened our doors in 1990. From White House trade officials moving into jobs with the foreign governments and corporations they once negotiated against to countries that abuse human rights procuring U.S. aid (with the help of Washington lobbying firms), from the unprecedented lobbying in 1994 on the issue of health care reform to *The Buying of the President* (Avon), our 1996 book about the special interests and "career patrons" behind the presidential candidates, the Center for Public Integrity's studies have investigated the national decision-making process and whether or not it has been distorted.

The attempts by the chemical industry to influence public opinion and public policy, via a prominent editorial and criticism from a respected third party, the American Cancer Society, days and weeks *before* an important news report, fascinated us. It is entirely legal, of course, not to mention understandable, for an economic interest to attempt to deflate the impact of anticipated public discourse about its products. What makes it relevant to us is that the public health is involved, and so is trust by the American people in their government. Even with today's crisis of confidence in politicians and government generally, most of us believe and assume that such agencies as the Environmental Protection Agency and the Food and Drug Administration exist, first and foremost, to safeguard us from harmful substances.

We decided to undertake a broad investigation of how chemical companies manufacture controversial products, year in and year out, in the face of government regulatory efforts, civil litigation by citizens who feel victimized, investigative news stories, congressional oversight of the federal regulatory apparatus, and other possible threats to the financial status quo. Here is some of what we found:

- Government at every level, but particularly in Washington, is overwhelmed by the efforts of an industry that is intent on keeping its products on the market. Both Congress and regulatory agencies have often placed a special interest—the chemical industry—ahead of the public interest they are charged with protecting. Many examples are documented in this book.

- Studies of the carcinogenic properties of the chemicals examined in this book—alachlor, atrazine, formaldehyde, and perchloro-ethylene—have varied in their findings, though all four chemicals have been shown to cause cancer in laboratory animals. Studies financed by the chemical industry have tended to find the chemicals innocent. Studies financed by nonindustry sources have tended to find them dangerous to human health. The industry-sponsored studies have provided enough of an argument to keep these and other chemicals on the market because the federal government's approach is to consider the chemicals safe unless they are proven to be harmful.

- We are always curious when we look at any government-regulated matter to find who has worked where. Of the 344 lobbyists and lawyers that we identified as having worked from 1990 to 1995 for the chemical companies and trade associations that this book explores in depth, at least 136 passed through what is commonly known as the revolving door, having previously worked for federal departments or agencies or in congressional offices. Virtually half of the EPA officials who left top-level jobs in toxics and pesticides during the past 15 years went to work for chemical companies, their trade associations, or their lobbying firms.

- Our investigation showed that EPA employees took at least 3,363 trips from March 1993 to March 1995 that were paid for—to the tune of $3 million—by corporations, universities, trade associations, labor unions, environmental organizations, and other nongovernmental sponsors. Four of the twelve corporations that this book examines closely—Ciba-Geigy, Dow, DuPont, and Monsanto—hosted EPA employees on at least 25 trips to their corporate headquarters and other locales. Organizations closely connected with these corporations financed many, many more trips.

Is it possible that the federal regulatory system, the way in which political campaigns are financed, the judicial system's increasing secrecy, the paucity of nonindustry funding for cancer research, and the news media's confusion about which scientist to believe *all* skew public discourse and policy in favor of the continued manufacture of fundamentally unhealthy products? The answer, after three years of intensive investigation, is yes. With millions—maybe even billions—of dollars to spend on lawyers, scientists, PR firms, campaign contributions, secrecy orders, and millions of pages and years of seemingly unlimited patience in litigation challenging the outmanned, underfunded government's every regulatory move, the chemical industry has managed to continue manufacturing what are generally considered to be harmful agents.

Choosing the four chemicals that this book examines in depth involved a process of consultation with 25 of the world's leading scientific experts on toxicity and cancer research. After selecting the four chemicals to be studied, we found two of the nation's best journalists on matters dealing with the environment. Dan Fagin, the environment writer at *Newsday*, was one of the three principal reporters whose coverage of breast cancer, including its links to pesticide use, qualified them as finalists for the 1994 Pulitzer Prize in explanatory writing. Before he began covering the environment in 1991, Fagin covered local, county, and state government in New York and Florida. Marianne Lavelle is a staff writer for the *National Law Journal*, where she has pioneered computer-assisted investigative reporting on racial injustice in U.S. environmental protection. Lavelle is the recipient of numerous awards for her investigative reporting on environmental issues, including the Scripps-Howard Foundation's Edward J. Meeman Award for Environmental Reporting, Columbia University's Paul Tobenkin Award, the George Polk Award for Legal Reporting, and the Women In Communications Clarion Award for Investigative Reporting.

Fagin and Lavelle were tireless in their reporting and writing on this project. They conducted approximately two hundred interviews and combed through tens of thousands of pages of documents, many of them internal government records obtained through the Freedom of Information Act. Amy Bohm and Robert Schlesinger, who directed and coordinated the research for this project over the course of three years, were equally tireless. As they have done before, members of the staff of the Center for Responsive Politics provided tremendous support and research help on finding the contributions made to Capitol Hill lawmakers by chemical companies, their allies, and their employees. Ellen Miller, the group's executive director, has our gratitude. Our efforts in finding documents, some of them obscure, were greatly assisted by the staff at Common Cause in Washington, and we thank Ann McBride, Jane Metzinger, and Katherine Loos for their help. Ken Niles of the National Library of Medicine provided invaluable guidance in finding countless journals and books.

The research for this study was underwritten principally by four foundations. We want to thank the Bauman Foundation, the Deer Creek Foundation, the W. Alton Jones Foundation, and the Giles and Elise Mead Foundation. The interest of these foundations in the environment is well known, and we appreciate their understanding why the project took so long and not interfering with the Center's independent investigation.

Finally, I want to take a few lines to explain why we do this kind of book. *Toxic Deception* is the 21st investigative study in the Center for Public Integrity's relatively short history, but it is one that explores an area vital to the public interest: not so much the potential or current threat posed by specific chemicals but the fact that government—the government of, by, and for the people—too often is complicit in ignoring, disdaining, and violating the interest of the people. In the case of the subject covered in the following pages, that abdication of responsibility could cost some citizens their lives. When we become aware of such things, we think it necessary to bring them to the public's attention. The hope, and there is some, is that citizens will use the information we provide to determine what they will do about it, how they will vote, what they will buy. We believe, and we show some examples of this in the book, that there are ways to make the interest of the people that which is held paramount. The first important step is to make sure that the relevant information is obtained and made known. We can't do it all, but we have tried to make the information we unearthed relevant and significant to understanding how government works—or, in some cases, doesn't work. Learning about the problem is part of finding the solution.

Charles Lewis
Executive Director
The Center for Public Integrity

Introduction:
The Invisible Threat

David Pinkerton liked to walk inside the skeleton of his family's home-in-progress on Orchard Road in Centralia, Missouri. Two or three nights a week, after his work as a lab technician at the University of Missouri, he and his wife, Mary, checked on the builder's progress. "It was our dream coming true," he would say years later. "The house was going to be ours forever."[1]

On one visit just a few weeks before moving day, David noticed a health warning printed on the subflooring that had been put in their new house. It noted that the product emitted vapors that could irritate the eyes and upper respiratory system, "especially in susceptible persons."[2] But David trusted the builder. "I figured he's putting this stuff in," Pinkerton recalled thinking. "He makes a living building houses. He wouldn't put anything in there that would hurt anybody."

After the new carpeting went in, David would never see the warning again. He left the house that evening feeling as if he had a cold. He chalked it up to the approach of winter and the stress of getting ready for the move.

The Pinkertons moved their bedroom furniture in first. "The heck with everything else," David said. "We were going to spend the first night in our new house." Soon they had put up their first big Christmas tree in the picture window. And they began to prepare for the December 30 wedding of Joy, the oldest of four daughters from Mary's previous marriage.

It was the second marriage, in fact, for both Mary and David. After

Joy left, there were five on Orchard Road: David's 13-year-old daughter, Jacinda; Mary's 13-year-old daughter, Brenda; and seven-year-old Kara, whom they had adopted.

Within a month, the three girls and their parents had grown quite ill.

David would sit in an old overstuffed chair until supper was ready; after dinner, he would usually go right to bed. At one point he missed two weeks of work. Mary could not cook or clean, and Joy would come to help. One night Mary tried to make dinner and David found her leaning against the wall with the skillet in her hand. Dirty clothes began piling up in the basement. All five had bouts of vomiting and diarrhea that would wake them up, almost nightly. Brenda no longer wanted to go to dance classes, even though ballet had been "her big thing in life," Mary later recalled.

"I had horrible headaches," Mary said. "I was tired. I had trouble understanding how I could feel so tired when I should be the happiest person in the world."

A friend invited the Pinkertons to stay at his house at Lake of the Ozarks, in central Missouri, in March. The weather was turning warmer. There was fresh air. Mary told David that she felt good there, better than she had for a long time. She had no headaches, so she started dusting her friend's home. "Hey, you came down here to relax and rest," David recalled their friend saying. "You didn't come down here to clean my house."

When they returned home on a Sunday evening, the Pinkertons had company, some friends from the First Baptist Church's youth group. Within an hour and a half, Mary was leaning on the counter in the kitchen. She felt so worn out that she could not even serve refreshments.

Later that evening, the Pinkertons began to suspect for the first time that the lingering and mysterious sicknesses had something to do with their brand-new home. Their suspicions were borne out a few months later when, at the Pinkertons' request, an inspector from the Missouri Department of Health came to their home. He attached a tiny glass tube to a portable pump and peered closely at the reading. The liquid in the tube turned purple all the way to the top of the scale, at the number 10: ten parts formaldehyde per million parts air—100 times the level at which many people become ill, according to the medical literature. The inspector's readings in other parts of the house showed at least 40 times the danger level.

The health inspector told the Pinkertons to leave the house, the very place they regarded as their dream come true. They were being poisoned slowly by formaldehyde gas.

The Pinkertons are just five of the thousands of people across the nation who fall victim each year to the invisible threats in every corner of the modern home, office, school, or yard. Few manage to trace their problems to one of the many chemical compounds that we breathe, eat, drink, and touch every day. But there has been incontrovertible evidence for some time that our daily use of a panoply of synthetic substances is helping to drive up rates of cancer, sterility, chronic fatigue, and many other diseases and illnesses.

The National Toxicology Program, which is operated by the National Institutes of Health, has designated more than 300 substances and processes as known or possible carcinogens, based on animal testing.[3] And while death rates have been declining because of better nutrition and medical care, more people than ever are contracting many of the cancers that are often associated with exposure to toxic chemicals, including multiple myeloma, non-Hodgkin's lymphoma, skin cancer, and cancers of the lung, prostate, bladder, brain, and breast.[4] Men born in the 1940s, in fact, have had twice as much cancer—and more than twice as much cancer not linked to smoking—as those born from 1888 to 1897, according to a study published in 1994 in the *Journal of the American Medical Association*. The same study found that women born in the 1940s, when compared to women born from 1888 to 1897, had a 50 percent higher cancer rate overall and a 30 percent higher rate for cancers not linked to smoking.[5] Some argue that we have much more cancer partly because people live longer and because we have eliminated many other diseases. But while the average lifespan has increased, age-adjusted analyses of cancer rates have shown that old age cannot account for the 44 percent increase in cancer incidence from 1950 to 1988.[6]

The National Cancer Institute, which has traditionally taken a skeptical view of environmental-related explanations for cancer, suggested recently that higher cancer rates are at least partly attributable to the proliferation of synthetic chemicals since World War II.[7] An article in its journal noted that job-related exposures to toxic chemicals may be responsible for increases in bladder cancer, that hormonally active pollutants may be driving up the incidence of hormone-induced breast tumors, and that some pesticides, hair dyes, and organic solvents may be partly responsible for the rising rates of non-Hodgkin's lymphoma.[8] Even pets are affected. Another study published in the *Journal of the National Cancer Institute* concluded that dogs are 30 percent more likely to get lymphoma if their owners use the weed-killer 2,4-D on their lawns and are more than twice as likely to get lymphoma if their owners apply 2,4-D more than four times a year.[9]

At least twenty peer-reviewed studies have linked various pesticides to cancer in children.[10] A recent study of 474 Denver children, for example, found that they were more than twice as likely to get leukemia if pest strips had been used in their homes and also were significantly more likely to get brain tumors or lymphomas if their homes had been treated by exterminators.[11] And two researchers at the National Cancer Institute, Aaron Blair and Shelia Hoar Zahm, who have conducted at least seven studies of farmers (a population that by most measures is healthier than the rest of us), have found unexpectedly high rates of leukemia, Hodgkin's disease, non-Hodgkin's lymphoma, multiple myeloma, and cancers of the bone, brain, connective tissue, eye, kidney, lip, pancreas, prostate, skin, stomach, and thyroid.[12]

"The patterns themselves suggest that there are some unexplained factors that have been widely introduced into the environment of industrial societies in the past several decades that account for these trends," Davis says. "We know about the farmers, that they die less often of lung cancer and heart disease but more often of non-Hodgkin's lymphoma, skin cancer, brain cancer and prostate cancer.... So we believe, my colleagues and I, that this argues for an environmental explanation."

Cancer is not the only serious health problem that has been linked to the proliferation of synthetic chemicals. Multiple chemical sensitivity and sick-building syndromes, though still controversial, have been recognized as conditions that are caused by exposure to toxic compounds.[13] Amazingly, indoor air pollution is now the most common topic of calls to the nationwide workplace-safety hotline operated by the National Institute of Occupational Safety and Health. Asthma rates in the United States have increased at least 30 percent since the 1970s, with the highest increases among children.[14]

Many studies of both humans and animals have found that synthetic chemicals can suppress the immune system by reducing the body's ability to produce antibodies and otherwise kill disease-carrying cells. Some research even suggests that exposure to man-made chemicals may be partly responsible for increases in infectious diseases, especially in children, the elderly, the chronically ill, and others whose immune systems are already weakened.[15]

Hormone-related health effects—especially infertility, breast cancer, and birth defects in the reproductive organs—are of special interest to researchers, because dozens of synthetic chemicals have been shown to perturb the intricate ebb and flow of estrogen, testosterone, and other hormones that guide sexual development in humans and animals. Experiments have shown that even relatively weak hormone disruptors, such as the pesticides dieldrin, endosulfan, and toxaphene, can produce particularly powerful effects when used in combination. A recent study of the effects of those three pesticides on estrogen-sensitive cells in test tubes, for example, found that the pesticides were 1,000 times more potent in combination than individually.[16] The research has especially disturbing implications because in the real world, of course, we all are exposed to a virtually infinite array of chemical combinations.

Today, according to the federal government, at least 70,000 chemicals are in commerce, with nearly six trillion pounds produced annually for plastics, glues, fuels, dyes, and other chemical products.[17] In 1995, the 100 largest U.S.-based chemical manufacturers sold more than $234 billion worth of chemical products—a 17 percent increase over the previous year—and made $35 billion in profits.[18] Their products have become such a pervasive part of American life, in fact, that an estimated 98 percent of all families now use pesticides at least once a year.[19] Every year, more than a billion pounds of pesticides are used in the United States—three-quarters of it on farms.[20]

Why are so many of these toxic chemicals still in our homes and on our lawns? The answer lies in the chemical industry's ability to manipulate the regulatory system to serve its own ends instead of the public's. Most Americans have been taught that the chemical revolution is nothing short of miraculous. Through the science and history films shown in school and the drumbeat of commercial messages at home, we've been told over and over again that we have chemicals to thank for our relatively carefree lifestyles. "Better things for better living" is the message of E. I. du Pont de Nemours and Company, while the catchphrase of Monsanto Company is "Without chemicals, life itself would be impossible."

But there is another side of the chemical revolution that few Americans think about until they—or people they know—are touched by disease or death. Until they have seen shipyard workers die with lungs crippled by asbestos dust. Or until they have watched soldiers carry home from Vietnam the sicknesses caused by their own army's chemical spray, Agent Orange. Or until they have witnessed young girls grow into womanhood sterile because of the synthetic estrogens doled out with abandon to their mothers.

It is not surprising that years after David Pinkerton and his family became ill, when they went public with their problems in a court of law, chemical companies pleaded with the judge in the case to make sure that the lawyers on the other side did not utter the word "asbestos" or the name of any of the other toxic substances that have made headlines in the past 30 years. A chemical manufacturer's success depends, of course, on its ability to maintain the patina of progress that is such a common theme in commercial messages.

When David Pinkerton stumbled on the printed warning on the subfloors of his new home, he could not have known that executives of a half-dozen mammoth corporations had spent five years arguing over those words and, ultimately, crafting them with care. Nor did he know how many millions of dollars they had spent on lawyers, scientists, public relations specialists, and lobbyists—all related, in one way or another, to the problem signaled by the warning notice.

The executives had woven a cloak of secrecy that hid the real dangers of their products as effectively as the carpeting covered the warnings beneath the feet of David, Mary, and the girls. Their companies—Borden, Inc., Celanese Corporation (now Hoechst Celanese), DuPont, Georgia-Pacific Corporation, and others—mapped out a strategy that has allowed them to continue selling dangerous, sometimes even lethal, products to millions of Americans every year.

The formaldehyde story is not unique. The same pattern of widespread exposure, documented health threats, and brazen industry manipulation appears again and again in the stories of the hundreds of toxic chemicals that are a virtually ubiquitous presence in our daily lives.

Just as the manufacturers of formaldehyde were able to keep its dangers hidden for so many years, so have the producers of many other toxic products managed to stave off—or at least blunt—efforts to crack down on dangerous products. They have done it by seeking to shape science, influence politicians, and buy publicity. They have alternately cajoled, soothed, and threatened the regulators who are supposed to protect the public. They have stirred up firestorms of "grass-roots'" support from farmers, small-business owners, and others who, ironically, are often hurt the most by the chemicals in question. Behind the scenes, all the while, their lawyers have sparred with tens of thousands of Americans who claim to have been injured by the chemicals—and, in many cases, negotiated secret settlements that keep the rest of us in blissful ignorance.

Manufacturers are not the only interests that try to manipulate the system. Environmental organizations, for example, sometimes misrepresent studies and employ other deceptive tactics to exaggerate the hazards of toxic chemicals. But their efforts are dwarfed by those of chemical manufacturers, who have the most at stake and the bankrolls to get the job done.

Consider the case of Benlate, a fungicide manufactured by DuPont. Although Benlate was supposed to kill molds while leaving plants alone, in 1990 thousands of farmers began reporting that the product was ruining their crops. Faced with a barrage of lawsuits alleging that Benlate was tainted with herbicides, DuPont launched a bare-knuckle campaign aimed at suppressing and discrediting scientific evidence that showed the fungicide was indeed contaminated.

Two secret company documents, eventually made public by judicial order, show just how far DuPont was willing to go.[21] A September 1991 letter written by one of DuPont's attorneys called for the company's scientists, who were supposed to be searching for the truth about Benlate, to instead become foot soldiers in DuPont's courtroom wars. "Scientifically, DuPont can maintain that it continues to search for a cause and that it will continue to do so as long as it appears necessary to address the issues raised by customers and regulator," Thomas Burke, an Orlando lawyer, wrote. "[Meanwhile,] in the litigation mode, we will not be forced into admitting that we have found a cause and it is our fault. It is a much better litigation position to state that we have looked, are looking, and will continue to look but have had no success, leaving the issue unresolved, than it is to have to admit that we have isolated the mechanism of injury."

The second DuPont document, written six months later, laid out the details of a secret plan—known internally as "Path Forward"—to fight the contamination claims by applying pressure "via legislators" on administrators at the University of Florida, where researchers were reporting results that contradicted DuPont's claims that Benlate was blameless. The secret plan also called for DuPont to build relationships with politicians and regulators and to move aggressively to undermine scientists who were critical of Benlate. "Cut them off publicly.... Don't share information with them.... Get intelligence on them so we're not blind-sided.... Know your enemies," the Path Forward document said.

However, at least four judges who presided over Benlate lawsuits would later conclude that at the same time DuPont was pushing its public line that Benlate was not contaminated, executives of the company knew otherwise. Noting that company scientists reported directly to DuPont's legal department, one of the judges said that DuPont's conduct "is suggestive of bias." Another fined DuPont $1.5 million for withholding damaging information, and the third slapped DuPont with a staggering $115 million fine for the same offense. The fourth judge didn't stop at a fine. Saying that the company showed "utter disregard" for ethics and legal procedure, she threw out DuPonts defense in a contamination case and awarded victory to a tree farmer.[22] DuPont is appealing all four rulings.

Despite the verdicts, the fines, and the bad publicity, DuPont's aggressive damage-control strategies are paying off. The University of Florida has virtually abandoned its Benlate research, prompting one scientist to retire in disgust. Without admitting guilt, DuPont has ended many of the lawsuits by reaching secret settlements in which the company has paid out more than a half-billion dollars and in return has insisted that farmers and their lawyers sign secrecy agreements and turn over potentially damaging documents. Most important of all, Benlate still has the unrestricted approval of the EPA, even though the agency regards it as a possible carcinogen because it causes liver tumors in lab animals.

DuPont's hardball campaign for Benlate is nothing new. Such efforts have been a staple of the chemical industry since 1962, when manufacturers worked to discredit Rachel Carson after the publication of *Silent Spring* launched the modern environmental movement. Back then, the chemical industry's efforts consisted chiefly of deploying friendly scientists to attack Carson in the press and distributing a sarcastic rebuttal to her book. Today, manufacturers are infinitely more sophisticated, carefully choreographing legions of lobbyists, lawyers, scientists, public relations experts, and mass-marketers for maximum effect.

This book is the story of how the chemical industry has managed to keep so many of its toxic products on the market, even in the face of mounting evidence of their danger and emerging—and safer—alternatives. It is also the story of how the federal agencies that are supposed to be the public's watchdogs have been defanged by the chemical industry's pressure tactics, which include junkets and job offers to government regulators, major contributions to politicians, scorched-earth courtroom strategies, and misleading multimillion-dollar advertising and public-relations campaigns.

To tell that story, *Toxic Deception* details the battles over formaldehyde and three other highly toxic chemicals that are in widespread use in the United States.* They are not the most dangerous, or even the most prevalent, chemicals on the market. But these four—all of which have been shown, beyond question, to cause cancer in laboratory animals—are emblematic of the thousands of toxic chemicals that are a pervasive yet often overlooked presence in the everyday lives of all Americans.

Atrazine, a farm weed-killer and the nation's most heavily used pesticide, is a health risk to consumers because it taints drinking water and because small amounts of it are present in corn, milk, beef, and other foods. Whether atrazine causes cancer in humans is still uncertain (the federal government classifies it as a possible human carcinogen), and we may never know for sure because researchers obviously cannot conduct controlled experiments on humans. But scientists can feed atrazine to laboratory animals, and those tests show that atrazine causes cancer in at least one type of rat—a fact that scientists have now known for 20 years. The recent discovery that the weed-killer interferes with the production of sex hormones may explain why laboratory rats have developed atrazine-induced tumors in their mammaries, ovaries, and uteruses. Based on the animal studies, the Environmental Protection Agency has estimated that midwestern corn farmers who mix and apply their own atrazine, get their water from a reservoir, and eat a typical diet face a one in 863 lifetime risk of developing cancer from atrazine; nonfarmers in the

*The four chemicals are among the most common in the United States and have a long history of government efforts at regulation. All were frequently cited as chemicals that pose significant health concerns in preliminary interviews that the Center for Public Integrity conducted in 1994 with scientists from a cross section of environmental, industrial, government, and research organizations.

Midwest face an estimated cancer risk of one in 20,747. By way of comparison, the EPA generally takes regulatory action when a chemical poses a lifetime cancer risk higher than one in a million. Yet Ciba-Geigy Corporation, the Swiss chemical giant that introduced atrazine in the United States in 1958, still sells about $170 million worth of the weed-killer every year, according to the National Center for Food and Agricultural Policy, a research group with close ties to agribusiness.

Alachlor, a frequent partner of atrazine in the weed-control arsenals of corn growers, is a risk to consumers and farmers for similar reasons. Best known by the trade name Lasso, alachlor shows up in water wells and reservoirs—and in food, too. The EPA has classified it as a probable human carcinogen, and the agency has estimated that people who get their water from midwestern reservoirs face a one in 250,000 lifetime risk of getting cancer from alachlor. High doses of the chemical have also caused liver degeneration, kidney disease, cataracts, and eye lesions in test animals. Worried about the health risks, a small but increasing number of farmers are employing innovative techniques to grow crops profitably with little or no use of chemical weed-killers. But most farmers never even hear about such innovations. Instead, in a springtime ritual that is as common in the Midwest as baseball games and church picnics, hundreds of thousands of farmers spray alachlor, atrazine, and other herbicides even before weeds begin to sprout. Unsurprisingly, alachlor still generates tens of millions of dollars in annual sales for Monsanto Company, and for many years it was the St. Louis–based agribusiness behemoth's top-selling pesticide.

Few chemicals have gained such intimate entry into our lives as **perchloroethylene**, the metal degreaser enlisted for use in the dry-cleaning industry in the 1950s. In theory, the solvent is safe within the confines of the tens of thousands of machines in which suits, dresses, and other clothes spin and tumble every day. In reality, it is not easy to contain any liquid as widely used as this one, which is manufactured in the United States by Dow Chemical Company, PPG Industries, Inc., and Vulcan Materials Company and in England by Imperial Chemical Industries Ltd. Perchloroethylene seeps and spills into groundwater, while its vapors invade nearby apartments and stores. And tests consistently show that consumers frequently bring "perc" home with them in their dry-cleaned clothes, particularly if their cleaners use old equipment or "short cycle" their machines—removing clothing before it is fully dried—to speed the process. Dry cleaners have always known that too much perc can make them dizzy or "drunk." There is a fierce debate over the effects of long-term exposure to smaller amounts of the chemical, but studies since the 1970s have linked perc not only to cancer but also to a variety of kidney, liver, neurological, and reproductive problems. The International Agency for Research on Cancer has classified perc as a probable human carcinogen, though it has settled on a determination that dry cleaning is only possibly carcinogenic.

Then there is **formaldehyde**, the chemical that invaded the Pinkerton home. A simple substance that actually is manufactured by all living cells in minute quantities, formaldehyde has become so ubiquitous in industry that it is impossible to catalog all of the products through which consumers might bring it into their homes. It is a preservative or binder in some cosmetics, pesticides, cleaners, and adhesives, though it is rarely labeled as such because it is not an "active" ingredient. To create permanent-press fabrics, the apparel industry coats textiles with a formaldehyde resin and bakes them. The primary use of formaldehyde—the use that this book explores in depth—has put it in the cabinets, flooring, walls, or furniture of virtually every American home built or renovated since the post–World War II housing boom. Nearly a dozen U.S. manufacturers produce strong formaldehyde resins, or glues, that are mixed with wood scraps and chips to produce particleboard, plywood, medium-density fiberboard, and other substitutes for solid wood. Few consumers realize they may be bringing formaldehyde into their homes when they install new kitchen cabinets or wood molding or buy furniture that they assemble themselves. Few buyers of mobile homes understand that they have a higher ratio of particleboard, and more formaldehyde gas, than conventional homes. The cheaper the wood, in fact, the more likely that it is a wood "product" held together with formaldehyde. Since the late 1970s, it has been known that formaldehyde causes cancer in rats, and therefore the International Agency for Research in Cancer terms it a probable carcinogen in humans. One of the most irritating gases to be cooked up in a chemistry lab, formaldehyde can gag, sicken, and weaken some people at even extremely low levels. In the worst cases, formaldehyde-sensitive people who have been exposed to high doses of the gas develop severe and debilitating asthmatic symptoms that appear whenever they breathe any chemical—from pesticides to perfumes.

David and Mary Pinkerton had spent their entire lives in the home state of Harry Truman, but in their case, after they discovered formaldehyde in their home, the buck did not stop anywhere. The builder referred them to a wood-products dealer, who advised them to fumigate their home with ammonia. By now, however, David Pinkerton had his doubts.

Georgia-Pacific and a smaller manufacturer, Temple Industries of Texas, would later hire scientists at big universities to say that the tests taken in the Pinkerton house were no good. They hired other testers to check the house—months later—and came out with much lower readings, around two-tenths of a part per million. And they brought other scientists to the witness stand to say that these levels of formaldehyde gas posed no danger.

The scientists conceded that formaldehyde can be "irritating to the eyes and upper respiratory system, especially in susceptible persons"— the words on the warnings underneath the Pinkertons' floor. As for the studies showing that rats that breathed formaldehyde developed cancer— well, they said, rats are not people.

When lawyers for Georgia-Pacific and Temple appeared before a jury in Clay County, Missouri, in 1989—five years after the Pinkertons had abandoned Orchard Street forever—they pointed out that the only federal safety standard for formaldehyde, established by the U.S. Department of Housing and Urban Development for manufactured housing, was no more than four-tenths of a part per million. The lawyers did not mention the industry's high-stakes lobbying campaign to get HUD to set a relatively high level and to block efforts by other federal agencies to regulate formaldehyde.

If the members of the jury did not like the law, the lawyer for Georgia-Pacific and Temple said, they should ask Congress to change it. The lawyer did not let on that his clients knew all about influence on Capitol Hill. The nation's major formaldehyde producers and users, for example, poured nearly $4 million into congressional campaigns from 1979 to 1995.[23]

The lawyer pointed out one more thing. It was about the Pinkertons. If they cared so much about their health, why did both of them smoke? Why was Mary obese, with high blood cholesterol? She had been ill many times in the past, had had her gall bladder removed, and had had a full hysterectomy. Mary's father was an alcoholic and so was her first husband. After her divorce, she had seen a psychiatrist.

The manufacturers had retained a clinical psychologist who rendered the opinion from the stand that Mary had a tendency to blame other people for her problems. It was his assessment that she had adopted a child to avoid closeness with her other children. As for the house on Orchard Road, the Pinkertons simply cared too much about it, said the psychologist hired by the company whose slogan was "See where your dreams and products from Georgia-Pacific can take you."[24]

At least several hundred families have confronted the manufacturers of formaldehyde as the Pinkertons did. But only a handful have ever agreed to stay the searing course of such litigation and to subject themselves, to being "peeled... back to the very core of their soul," as the Pinkerton's lawyer, R. Frederick Walters of Kansas City, put it as their trial drew to a close. Most plaintiffs agree to cash settlements, in which the chemical manufacturer admits no harm, and they promise not to reveal the money they garnered, their stories, or the documents they gathered to bolster their cases.

But the Pinkertons persevered through the longest trial in the history of Clay County. On December 21, 1989, an outraged jury awarded them $140,000 for medical expenses, and $63,000 for property losses; on January 8, 1990, the jury added $16 million in punitive damages. Their story is one of the few on the public record.

Most of their story, that is.

Ever since her six months on Orchard Street, Mary Pinkerton, who had never been very healthy, suffered from a condition known as multiple chemical sensitivity. She grew ill with asthmatic symptoms whenever she was near any chemicals, not just formaldehyde. A year and a half after the verdict, on a family trip, the Pinkertons stayed in a hotel that they later learned had been sprayed with insecticides. Mary became unable to breathe and died on July 1, 1991.

"You know, everybody walks through life and some people carry heavier burdens than others," Walters, the lawyer for the Pinkertons, had said at the trial. "I'm here to tell you [that Mary's] not a strong person. She has had previous problems. But [Georgia-Pacific and Temple] have added to the burden.

"Mary walked into that house thinking her life was going to change, things are going to get better. And six months later, her burden is crushing. You know what? She's never going to get through her burden now."

1

Presumed Innocent

It began with bugs. Insect-borne diseases were killing even more American soldiers than the Axis armies in 1943 when J. R. Geigy SA., a 185-year-old Swiss dye-making firm, applied for a patent on a miraculous new insecticide, dichlorodiphenyl-trichloroethane—DDT for short. DDT was less toxic to humans than the crude lead-arsenate insecticides then in use, but it could kill hundreds of species of pests and stayed lethal long after it was sprayed. The U.S. government immediately declared DDT vital to the war effort and arranged to have millions of soldiers and refugees "dusted" to prevent malaria and typhus.

With the end of the war, sales of DDT scaled even greater heights as farmers, public health officials, and the rest of America giddily embraced "the miracle white powder." A 1947 advertisement in *Time* magazine featured dancing vegetables, animals, and a farmer's wife singing "DDT is good for me-e-e!" The following year, *Life* magazine breathlessly touted DDT's benefits by picturing a swimsuit-clad teenage girl eating a hot dog in a cloud of DDT fog. That same year, 1948, Geigy's Paul Mueller was awarded a Nobel Prize—in medicine—for having discovered DDT's insecticidal properties almost 10 years earlier. Awarding Mueller the prize in the category of medicine later proved an ironic choice, considering that DDT would be banned in the United States in 1972 as a possible carcinogen and threat to wildlife.

The chemical revolution had begun. By 1951, Geigy and its

1

licensees, including DuPont, General Chemical Company, and Hercules, Inc., were selling a staggering $110 million worth of DDT a year, up from $10 million in 1944, and a new industry was racing to synthesize the next miracle powder.[1] By reshaping molecules into an endless variety of configurations that do not exist in the natural world, chemists created the products that perform the dirty work of industrial progress: solvents that strip away dirt and grime, a seemingly limitless array of plastics and resins, pesticides that cripple the nervous systems of insects and terminate photosynthesis in weeds. Molecule by molecule, company scientists built the "better things for better living" that the nation's second-largest chemical company, DuPont, touts in its television commercials.

But there is a dark side to the chemical revolution, the dim outlines of which have only gradually started to come into focus in the half-century since DDT was patented by Geigy, a predecessor firm to the giant company that is now known as Ciba-Geigy Corporation.[2] It turns out that the same qualities that make synthetic chemicals so valuable— their potency and durability, for instance—also make them dangerous. Many of them stay in their original forms for months, years, or even centuries—more than enough time for them to make their way into the air we breathe, the water we drink, and the food we eat.

The chemical industry argues that its products pose no cause for concern. The industry line is this: Yes, toxic chemicals can be dangerous when used improperly, but, when they are handled with care, people are exposed to them only in tiny amounts. The big chemical manufacturers, in fact, have a program called Responsible Care, the emblem of which is two hands cradling a molecule. The industry frequently points out that plants create toxins to ward off pests and natural predators. The amount of these natural cancer-causing chemicals in the daily diet is 10,000 times greater than the residue of pesticides and pollutants created by man, according to a 1996 study by the National Academy of Sciences. The study was financed partly by the federal government and partly by a group of pesticide and food companies that included Ciba-Geigy, Dow, and Monsanto.[3] It echoes the industry's long-held position that natural carcinogens work in the same way that synthetic chemicals do and that one is no worse than the other.

Not everyone agrees with this assessment, however. Another wing of the scientific community points out that natural toxins are rarely deadly at the levels found in nature. Lead, for example, occurs naturally but is a major health threat only because it is concentrated in such man-made products as paints and pipes and is also in smokestack pollution from garbage incinerators. These scientists also argue that man-made chemicals pose a special threat because our bodies are ill prepared to break them down into harmless compounds. Humans have had millions of years to evolve ways of metabolizing naturally occurring toxins. We are badly overmatched by synthetic chemicals that typically are only a generation or two old, a mere eyeblink on the time scale of human evolution. That is why DDT, a quarter-century after it was banned, is still in the blood of virtually all Americans, even in children who have inherited it through the placenta and breast milk.

"We have had no time to develop defense mechanisms,'" says Carlos Sonnenschein, a professor of cell biology at the Tufts University School of Medicine. "Perhaps in another hundred million years we will be able to adapt and overcome the effects of these chemicals—if we can make it until then."

We are all, in effect, involuntary participants in a vast, uncontrolled experiment. We have enjoyed the benefits of man-made chemicals, but we have unknowingly taken a great risk. We have unleashed tens of thousands of man-made chemicals into the air, water, and soil without knowing what the effects will be on the world's living creatures, including *Homo sapiens*.

"It's a Faustian bargain we have made," says Devra Lee Davis, a senior fellow at the World Resources Institute, a former senior adviser to the assistant secretary for health in the Department of Health and Human Services, and one of the nation's top experts on cancer trends. "I don't think that anyone who has lost a breast or testes, or can't have children, would agree that it's been worth it."

Davis is referring to some of the newest research on chemical toxicity, which is chronicled in the 1996 book *Our Stolen Future*. The book's coauthor, Theo Colburn, a senior scientist with the World Wildlife Fund, has amassed research showing that many chemicals we have set loose on the environment appear to mimic estrogen in the bodies of humans and animals, disrupting the delicate balance vital to the proper functioning of their immune, nervous, and reproductive systems. Col-burn's work traces the rise in sexual and other developmental defects and malfunctions in polar bears, sea gulls, and other species across the world and behavioral deficits in children born to women who were exposed to high levels of chemicals from eating contaminated fish.[4]

Add these concerns to the traditional worries about cancer and one must ask: Are *all* chemicals dangerous?

Obviously not, since life itself is based on chemicals and their interactions. More than 99 percent of the earth's crust, oceans, and atmosphere consist of just 10 elemental chemicals: oxygen, silicon, a handful of metals and minerals, and hydrogen. And, of course, the human body is 93 percent oxygen, carbon, and hydrogen.[5]

But many man-made chemicals—and the ways in which they are used—raise special concerns. Most of the chemicals that are widely manufactured are either man-made versions of chemicals that appear in tiny amounts in nature or brand-new combinations of these uncommon natural chemicals. Chlorine, for example, makes up less than 0.2 percent of all chemicals in nature, yet about 77 million pounds of chemicals containing chlorine are now manufactured in the United States every year.[6]

Unfortunately, the very characteristics that make some chemicals so valued in the commercial world are the same characteristics that make them potentially harmful to people and the environment. Asbestos, of course, was prized for its durability and resistance to flames. The fibers, indeed, proved stronger than the lungs of thousands of people who worked with it. Chlorinated chemicals are solvents and pesticides that do not catch fire and do not lose their potency for a long time—so long, in fact, that DDT continues to accumulate to alarming levels in the fatty tissues of Great Lakes fish nearly a generation after its use was banned in the United States.

Both asbestos and DDT are off the U.S. market. Wouldn't the government likewise pull the plug on any other substances that are known to be dangerous? Not necessarily.

The Environmental Protection Agency, with a $6.5 billion budget and 19,000 employees in 1996, is the largest bureaucracy in the federal government that is devoted solely to regulation. The agency's regulations have put pollution-control devices in American automobiles and forced hundreds of thousands of factories across the United States to control and monitor how much pollution they release into the environment each year. The EPA has excavated abandoned pits where manufacturers dumped poisons for generations; in Love Canal, New York, Times Beach, Missouri, and a half-dozen other places, the agency moved whole towns' populations after judging the residents to be at severe health risk from pollution.

The EPA's record of success is substantial. Blood lead levels in the United States have plummeted. Fish and wildlife have returned to the Great Lakes. But these and other achievements came only after years of struggle, lawsuits by both environmentalists and industry, and compromise.

The EPA was born in 1970 of an uneasy marriage of 1960s social activism and Nixon administration domestic policy. Demonstrations and teach-ins on April 22, 1970, known as Earth Day, highlighted public outrage at dead lakes and rivers and thick city smog. Senator Edmund Muskie of Maine, then considered one of the strongest Democratic challengers to Richard Nixon, had gained nationwide popularity with his environmental advocacy. Nixon countered with his own proposal: to create one big Environmental Protection Agency by bringing under one roof 15 small agencies that already existed in other federal departments.

This fragmented beginning has had a lasting effect on the way the EPA does its job. In 1996, the EPA still had separate offices to oversee problems of water pollution, air pollution, waste disposal, and toxic substances, even though these problems are interrelated. The problem of toxic substances really is a water, air, *and* waste issue, all at once. Naturally, turf battles frequently erupt that prevent the EPA from effectively protecting the public—as when the Office of Air and Radiation and the Office of Toxic Substances and Pollution Prevention both were moving in separate directions in the early 1990s on what to do about the dry-cleaning chemical perchloroethylene.

Because few other agencies have such profound potential to affect the way American business operates, the EPA has spent more than its fair share of time being a political football. President Ronald Reagan drastically cut its budget and workforce in his first year in office. Congress not only restored money and staff in the following years but also gave the EPA responsibility for huge new environmental programs. As recently as 1995, the EPA was operating at a de facto 25 percent budget cut because of a dispute between the Republican-controlled Congress and President Bill Clinton over how much money and authority the agency should have.

The combination of budgetary uncertainty, unwieldy design, and the conflict inherent in attempting to protect both the public and business from unreasonable burden means that the EPA does its job in an atmosphere of considerable turmoil.

It never has been easy for the agency to take action against a chemical that it suspects is hazardous. We now know that lead is harmful, especially to young, developing minds, but for more than a decade the oil industry fought the EPA's efforts to phase out its use in gasoline. Lawsuits, not the federal government, ended most uses of asbestos; until the day they were driven into bankruptcy, its manufacturers argued that it was not hazardous when used properly.

Many chemicals remain on the market today that are suspected to be dangerous. But as they did with asbestos, lead, and even DDT, manufacturers raise doubts about these indications of risks. Suspect chemicals remain on the market because we do not know *for sure*.

Everyone probably has one or more of these chemicals at home. Monsanto Company's Ortho Weed-B-Gon Lawn Weed Killer is labeled with its active ingredient, as the government requires: sodium 2,4-dichlorophenozyacetate. What is not on the label is that this popular weed-killer, known as 2,4-D, is also suspected of being carcinogenic and otherwise toxic to the neurological and reproductive systems. Alberto-Culver USA, Inc., informs consumers on the label of its VO5 hair conditioner that it contains formaldehyde and FD&C Red #4. But how many people know about the potential carcinogenicity of these two chemicals? How many know that formaldehyde also has been shown to cause dermatitis and other skin problems in some people?

Stories involving these and other suspect chemicals make news every day. A seventy-six-year-old man in McKeesport, Pennsylvania, dies in June 1994 while working with 2,4-D and other pesticides.[7] A 30-year-old navy lieutenant develops a rash, then basketball-sized blisters, then internal organ damage, and dies in 1982 after a game of golf on a course that had just been sprayed with the fungicide Daconil 2787, which is still on the market.[8] A 53-year-old Texas woman who works in a nail salon—nail salons are typically saturated with chemical fumes—reports muscle problems, chronic fatigue, liver damage, bronchial asthma, chronic diarrhea, and a slew of chemical sensitivities.[9] Since 1991, thousands of veterans of Operation Desert Storm have complained of lethargy, diarrhea, numbness, memory loss, and sleep disturbances; research in

1996 revealed that such symptoms could very well be caused when the common insect repellents DEET and permethrin are combined with the anti-nerve-gas pill pyridostigmine—a chemical cocktail that soldiers were exposed to for their own protection.[10]

The manufacturers of the four chemicals that this book examines in depth—alachlor, atrazine, formaldehyde, and perchloroethylene—all argue that there is not enough evidence of risk. They have succeeded in keeping their products on the market with one line of defense: We don't know *for sure* how risky these chemicals really are.

"We use scientific uncertainty like a cross to the vampire," Davis says. "I readily concur that there are major uncertainties about the relationship between chemicals in the environment and human health, and as a scientist I'm working to resolve them. But as a citizen, and a mom, I am concerned about what we know already, and I want to take prudent precautions."

The warning signs have been visible for more than 200 years. The links between worker disease and exposure to various industrial residues, chemicals, and cleaners date to 1775, when Percivall Pott published his studies of scrotum cancer in English chimney sweeps, who labored in soot and coal tar.[11] Not long after the publication in 1962 of Rachel Carson's *Silent Spring*, the tiny Japanese fishing village of Minamata became known worldwide as an environmental horror. Nearly a thousand villagers suffered excruciating deaths and thousands more the lingering effects of being poisoned by the tons of mercury-based compounds that Chisso Corporation had dumped into one of Japan's most productive fishing grounds.

Then there is an indisputable fact of modern life: Today, cancer kills one of every five Americans. A poignant illustration of the power and prevalence of the disease came at the start of the Pinkerton formaldehyde-injury trial in Missouri, when the family's attorney, Fred Walters, began asking potential jurors some questions.

He asked whether any of them, or any of their family members, had suffered any "latent injuries." No one raised a hand. "It's something you can't see," he said, "a latent disease, problem, or dysfunction." More silence.

Then Walters tried another approach. "Is there anyone here who has had a loved one who has experienced cancer?" Thirty-six hands went up. He began to interview each person.

Skin cancer, breast cancer, lung cancer, leukemia. Brain cancer, colon cancer, kidney cancer. "A vast tumor in the sinus region." Bone cancer. "Cancer of the jaw that started in his back and went all over," one said. Another explained, "My mother died of ovarian cancer."

"My grandfather died of stomach cancer," another woman said. "I would also like to respond to the previous question about latent disease. My mother took DES [diethylstilbestrol, an anti-miscarriage drug containing a man-made form of natural estrogen that was prescribed briefly in the late 1950s, before its dangers were known]. Both my sister and I have some chance of cancer."

Another told of his uncle, who had been mustard-gassed in World War I and died of asbestos-caused cancer. Yet another told of having to move out of a house after it had been treated for termites with the pesticide chlordane because the fumes were so strong. "My mother died two years ago of a brain tumor," that person said.

Bring together any large group of people and you will hear similar stories. Public-health officials believe that cancer soon will surpass heart disease as the nation's number-one killer. Are environmental exposures— to such things as termite-killers, drugs, and the formaldehyde that this jury would learn about over the weeks of the trial—responsible for some forms of this deadly illness?

The world has been slow to grasp the dimensions of the health problems caused by toxic chemicals because the evidence has often been so difficult to tease out. To know where something is, you have to know where it isn't. Yet scientists have found it extremely difficult to separate the world into exposed and unexposed populations. We are all exposed to so many different chemicals, in fact, that it is difficult to pinpoint the effects of any particular compound. Linking specific chemicals to specific health problems is also extraordinarily difficult because there are so many other causes of cancer, respiratory disease, genetic damage, immune deficiencies, and rashes—to name just a few of the conditions that can be triggered by chemical exposures. It is rare indeed to find a disease that is caused only by one—and by only one—toxic compound. One such disease, asbestosis, is triggered by asbestos fibers lodging in the lungs. When a doctor diagnoses asbestosis, there is no doubt as to the cause. But that is never the case with cancer, for instance, or infertility. A final complication is that women, young children, and the elderly tend to be more susceptible to chemical-induced illnesses than adult males, for reasons that scientists still do not fully understand.

Still, evidence has been accumulating that many man-made chemicals harm people. Acetone, an acutely toxic chemical that causes cancer in laboratory animals, is a key ingredient in a panoply of products, ranging from nail-polish remover to paint thinner. Vinyl chloride is in all kinds of plastic products—including PVC water pipes—despite scientific evidence that it can cause liver, brain, and lung cancer in humans. The solvent methyl chloroform, also known as 1,1,1-trichloroethane, is still in many kinds of carpeting, spot removers, and glues, even though it is so acutely toxic that breathing high quantities can be fatal. Many of the most common lawn pesticides used in America, including the weed-killer 2,4-D, the fungicide chlorothalonil, and the insecticide chlorpyrifos (Dursban), have been shown to cause cancer or birth defects in lab animals. The list goes on and on.

In all, about 300 man-made chemicals—many of them still in wide use—have been identified as cancer-causing agents in animal tests conducted by researchers in the National Toxicology Program of the National Institutes of Health. The true number of carcinogenic chemicals is probably far higher, since the program has not conducted even preliminary screenings on more than 80 percent of the chemicals currently on the market. And as scientists increasingly expand their research beyond the traditional focus on cancer, they are learning from laboratory tests that many chemicals can also wreak havoc in the body's immune and reproductive systems, triggering birth defects and sterility and leaving victims open to deadly infectious diseases.

Just how real is the threat? Conducting experiments on humans is unthinkable, and tracing a case of cancer or a birth defect back to a specific toxic exposure is virtually impossible. Consequently, some of the strongest scientific evidence is based on tests in which a few hundred lab animals are fed high doses of a chemical for a year or two in an effort to mimic the chemical's effects on the millions of people who may be exposed to it at much lower levels over a lifetime.

But not all the scientific data comes from rats or rabbits. Groups as diverse as beauticians, veterans of the Persian Gulf war, farmers, and factory workers have complained of health problems they attributed to chemical exposures. By using the tools of epidemiology—the study of disease rates in broad populations—researchers have bolstered the claims of some of those groups and confirmed the validity of animal testing. Epidemiologists at the National Cancer Institute, for example, have discovered that farmers have unusually high rates of non-Hodgkin's lymphoma, brain cancer, and several other diseases that have been linked to certain pesticides in animal studies. Other studies have found that children are more than twice as likely to get leukemia if pest strips have been used in their homes and that dogs are 30 percent more likely to get lymphoma if their owners spray 2,4-D on their lawns.

The animal tests and epidemiologic studies collectively form a powerful case that man-made chemicals are indeed taking a major toll on human health—and on the economy, too. By one widely quoted estimate, made in 1989 by the World Health Organization, pesticides alone are responsible every year for one million poisonings and 20,000 deaths worldwide. Every year, the use of pesticides costs the U.S. economy more than $12 billion, with much of the cost attributable to pesticide purchases, medical costs, water pollution, and the poisoning of animals, birds, and beneficial insects, according to a study by David Pimentel, a researcher at Cornell University.[12] And those estimates of deaths and economic costs do not include the damage caused by the thousands of nonpesticide chemicals that are in everything from paint to perfume. Pimentel is no abolitionist. In the United States, he estimates, the use of pesticides averts about $16 billion in crop damage a year. The problem, he says, is that toxic chemicals are used reflexively without even considering other options that save lives and money.

In the midst of this litany of man-made risks, it makes sense to ask about the role of the federal government in regulating toxic chemicals. What about the Environmental Protection Agency, the Food and Drug Administration, the Consumer Product Safety Commission, and the Occupational Safety and Health Administration? Surely no chemical could get on the market—let alone stay there—if it were not safe.

This book is the story of why none of us should be so sure.

The federal government does not screen chemicals for safety before they go onto the market. Why not? For the most part, it is because the chemical industry was a firmly entrenched economic and political force by the time that Congress was moved to do something to protect the public from hazardous compounds in the 1970s.

As a result, there is a stark contrast between the EPA's powers to regulate chemicals and the powers of, say, the Food and Drug Administration. The FDA was born in the 1920s, before many of today's pharmaceutical products were even imaginable. It has always had the authority to require that a drug be proven both effective and safe before it is put on the market in the United States. The agency may have neglected to use its authority at times, and it certainly has made mistakes over the years and allowed dangerous drugs to reach the public. But these are outweighed by the stories of its action to keep such drugs as thalidomide, which caused so many birth defects in England in the 1960s, off the market in the United States.

In the 1970s, however, every environmental protection law was forged in the cauldron of hot debate over whether the federal government's job was to protect people or to protect a thriving and well-established industry. Each law reflected a compromise between those two positions.

Pesticides would have to be "registered"—certified by the federal government as posing no unreasonable risk to health or to the environment. But the bug- and weed-killing compounds could stay on the market while the Environmental Protection Agency completed the assessments, and the assessments would be based on studies that the manufacturers themselves supplied. The built-in incentives for delay are obvious: If a product is hazardous, the longer a manufacturer takes to provide the EPA with data, the longer the product stays on the market.

Indeed, the EPA has formally registered only about 150 pesticides. Thousands more are on the market awaiting review. "Most of these products may continue to be sold and distributed even though knowledge of their health and environmental effects is incomplete," the General Accounting Office, the investigative and watchdog agency, said in a May 1995 report to Congress.[13]

As for other toxic chemicals, the EPA's Office of Pollution Prevention and Toxics is responsible for deciding which pose a "substantial risk" to health and whether they should be controlled or removed from the market. But Congress again sided with the chemical industry when it passed the Toxic Substances Control Act in 1976. Chemicals would stay on the market unless they were proven to be a risk, and the EPA would not conduct its own safety tests but would instead rely on research submitted by the manufacturers.

The EPA must rely on such studies because it has never been given the money, staff, or legal authority to do anything else. By 1994, in fact, the EPA had issued regulations to control only nine chemicals in 18 years, according to the same grim report by the General Accounting Office.

In sum, only a fraction of the 70,000 chemical compounds sold today have been examined for safety, and most safety tests are conducted by manufacturers *after* questions have been raised about a product.

Government regulators plead that they have neither the manpower nor the money to handle consumer protection any other way. But that ignores the other side of the equation. The scales have been tipped by corporate money and clout.

As a result, even as science uncovers such new dangers as the hormone-disruption threat, the regulations that are supposed to protect Americans from unsafe chemicals remain mired in the gee-whiz era of "DDT is good for me-e-e!" Most of the EPA's pesticide regulations, for example, still assume that healthy adults are the only people at risk from herbicides and insecticides, despite the mounting evidence that infants and children are much more vulnerable.[14]

Indeed, the most important law governing toxic chemicals and pesticides has its roots in the postwar years when government officials focused on promoting the U.S. chemical industry. Adopted without controversy in 1947, the Federal Insecticide, Fungicide, and Rodenticide Act (FIFRA) was designed not to assure the public that pesticides were safe but to assure farmers that the products were sufficiently lethal against insects and weeds. The nation's chemical industry, whose representatives had helped to draft the law, backed it enthusiastically as a way to keep fly-by-night firms out of the industry and to ensure that friendly federal regulators would be in charge instead of 50 states with 50 sets of regulations. The new law included only the scantest oversight of the chemical companies and required merely that manufacturers register their products with the federal government and back up claims of safety and effectiveness on product labels.

When the law was amended after the dawn of modern environmentalism in 1972, the chemical industry was again in the driver's seat. Instead of requiring that manufacturers prove their products to be safe, Congress put the burden on the government to prove that they are dangerous. In language that remains in place today, lawmakers told the newly created EPA that it could not take action against a pesticide unless it could prove "an unreasonable risk to man or the environment, taking into account the economic, social, and environmental costs and benefits of the use of any pesticide."[15] That is an especially important edge for manufacturers, because there is so much that scientists do not know about the way chemicals behave in the human body and in the environment. Even when researchers are able to show that a chemical causes tumors in animals and can trace death and disease in humans back to the presence of the chemical, they almost never can fully explain why or how this happens, or why some people are susceptible and others are not. Amid so much scientific uncertainty, manufacturers find it easy to raise doubts about any research that links adverse effects to their products.

Regulators often end up making no decision as to whether a chemical should be withdrawn from the market. That is a perfectly acceptable result to manufacturers; for them, because chemicals are deemed safe until proven otherwise, a tie is as good as a win.

Such a pro-industry bias is the rule, not the exception, in federal regulation of toxic chemicals. When the Toxic Substances Control Act was passed in 1976, it seemed, on its face, to give the federal government enormous power to protect the public from dangerous products. But the law stipulated that the EPA must weigh the likely costs of its decisions against the potential benefits and prove that it had chosen to use its power in "the least burdensome" way to industry.[16] Consequently, the EPA has been able to regulate only a relative handful of chemicals. Even its ban on products containing asbestos, a known carcinogen, was overturned by a federal appeals court.[17]

Federal environmental laws not only tend to be weak, but they also frequently ignore the biggest health risks; otherwise, major restrictions on the sale of toxic products would be required. The environmental laws of the 1970s and 1980s focused on cleaning up pollution in lakes, rivers, and streams, in the air, and at abandoned industrial waste sites—in some cases, with great success. But the EPA estimates that the *use* of toxic products accounts for 50 to 90 percent of the cancer risks from air pollution, while the *disposal* of toxins—in landfills and incinerators—accounts for only 3 to 25 percent.

At the most fundamental level, the federal regulatory system is driven by the economic imperatives of the chemical manufacturers—to expand markets and profits—and not by its mandate to protect public health. The EPA and other federal agencies, for example, have thousands of employees—scientists, lawyers, and others—who are responsible for administering complicated regulations under FIFRA, the Toxic Substances Control Act, and other laws that require elaborate testing to sniff out the potential dangers of synthetic chemicals. But those same laws also require that the tests be conducted by or for the parties that have the least reason to be objective: the chemical manufacturers.

Fifty years after the chemical revolution began, there is a vast library of literature about the toxic effects of synthetic chemicals. There are studies by institutes, universities, agencies, corporations, advocacy groups, and many others. But in the eyes of the EPA, most of the shelves of that library might as well be empty, and the few books available might as well all be from the same publisher, articulating the same point of view, because the agency usually relies on research conducted by or for manufacturers when it is time to make a decision about regulating a toxic chemical.

There are several reasons for that. The first one is purely practical: It costs the EPA a lot less to have someone else to do the research. "Absent Congress and the general public funding the Environmental Protection Agency's programs with sufficient resources to do independent research,...there are few alternatives out there to doing business this way," says Daniel Barolo, who heads the EPA's pesticides office. "We don't apologize for the lack of resources. What we try to put in place are professionals and procedures that will achieve the most independent evaluations possible of the studies that are given to us."

Critics of the agency argue that it often falls short. "The EPA is just a paperwork check," Patti Goldman, a lawyer with the Sierra Club Legal Defense Fund, says. "Its really a self-regulating system. The industry does the tests and presents the results."

It might be more accurate to say that the EPA *hopes* the industry presents the results. One sign that the government's trust may be misplaced came in 1991 and 1992 when the EPA offered amnesty from big-money fines to any manufacturers that turned in health studies they should have provided to the agency earlier. Manufacturers suddenly produced *more than 10,000 studies* showing that chemicals already on the market could pose a "substantial risk of injury to health or to the environment"—the kind of never-published data that the law says must be presented to the government immediately.

The fact is that no one cares more about the fate of a chemical than its manufacturer. Industry officials are a near-constant presence within the agency, exerting pressure in ways that environmental groups and independent scientists simply cannot match. "It's democracy in action, only it's only one side that's in action," says Richard Wiles, a former senior staff officer of the National Academy of Sciences Board on Agriculture who is now the vice president for research at the Environmental Working Group.

In the case of the withheld studies, for example, the EPA was so overwhelmed by the deluge of documents, some of them dating back to the 1960s, that its staff named the computerized system for keeping track of them the "Triage Database." At first, the EPA's staff tried to characterize the studies as triggering a "high," "moderate," or "low" level of concern. But it has never been able to complete the job, and the mountain of studies remains part of the agency's enormous backlog on assessing chemical risks. The 10,000 studies indicating "substantial risk" that manufacturers had failed to submit to the EPA included at least three that dealt with alachlor, 22 that dealt with atrazine, 35 that dealt with formaldehyde, and six that dealt with perchloroethylene.

Many of the documents look as if they came out of the Central Intelligence Agency instead of corporate science labs. Monsanto, for example, stamped its studies COMPANY SANITIZED and blanked out entire passages, placing empty brackets around the redactions.

When Congress wrote the Toxic Substances Control Act in 1976, it gave manufacturers broad permission to claim certain information as trade secrets. The EPA has sparred repeatedly with manufacturers over the years about the amount of data that they keep secret as "confidential business information." The maufacturers have won hands down. On July 15, 1992, for example, taking advantage of the amnesty program, Monsanto turned in to the EPA a study on a chemical mixture that it identified only as "developmental alachlor formulation TV."[18] All that can be gleaned from the document is that the mixture contained at least five chemicals that are on the government's list of potentially hazardous chemicals and that it appeared to cause skin sensitization in guinea pigs. And some of the chemical-risk studies belatedly submitted to the EPA dealt with issues that have only recently emerged in the public consciousness—the reproductive effects of chemicals, for example. On December 8, 1992, the Halogenated Solvents Industry Alliance submitted data showing that the offspring of rats exposed to high doses of perchloroethylene had lower body weight at birth than the litters of unexposed rats.[19]

The information that the manufacturers present is sketchy and full of spin. The industry alliance makes much of the fact that the stunted perc rats appeared to plump up over time to the same size as the unexposed rats. But this focus glosses over the real reason that toxicologists care about low birth weight: In human infants, low birth weight can signal physical health, behavioral, or intelligence deficits that cannot be resotred by gaining weight.

Regulators and industry scientists talk to one another in an arcane scientific language that few others understand. They are specialists who share a professional kinship that outsiders cannot penetrate. It is no wonder, then, that EPA officials so often seem to be most comfortable with research generated by chemical companies.

"The people who have a vested interest in continued dependence on pesticides are the people to whom the EPA looks to find answers to regulatory questions, and that's precisely why we don't see the changes that we ought to be seeing," says Jay Feldman, the national coordinator of the National Coalition Against the Misuse of Pesticides, an activist group.

In short, chemical manufacturers have learned through experience that they can keep dangerous products on the market as long as they harp on scientific uncertainty and stay in control through every stage of the process: in the laboratories where safety data are generated, in reports in which the results are presented to regulators, in the courts, and in the arena of public opinion. More often than not, the chemical companies succeed brilliantly.

Born in war, the chemical revolution rolls on, its troops overpowering any obstacles in their path. Here's how they do it.

2

Four Chemicals

Atrazine

The white crystalline powder might as well have been magic. Sprayed onto a broad variety of plants, the newly synthesized chemical with the tongue-twisting name of 2-chloro-4-ethylamino-6-isopropylamino-1,3,5-triazene seemed to have eyes. It quickly killed grasses and weeds but left unmolested corn, cotton, and other big-money crops. The scientists at J. R. Geigy SA who had synthesized it in Switzerland in 1955 may not have been sure at first what they had created, but within a few months, after the first field tests in the company's greenhouses, they knew that they had hit the jackpot. The chemical would be called atrazine.[1]

The scientists knew all about miracle chemicals. A dozen years earlier, Geigy had patented an insecticide known as DDT, and ever since then the company had been looking for a chemical that could do to weeds what DDT did to insects. The tests in the greenhouses at Basel suggested that the compound that Enrico Kneusli, a Geigy chemist, had first created on July 26, 1955, was indeed the second coming of DDT.

Atrazine's magic was grounded in its quirky chemical structure. Like other herbicides, it could kill plants quickly by stopping photosynthesis, the process by which plants convert carbon dioxide and light into food. Unlike the other popular weed-killers of the day, however, atrazine worked effectively against both grasses and broadleaf weeds, the twin banes of corn farmers. But what was truly innovative about atrazine was

17

that corn, sorghum, sugarcane, and a few other popular crops are naturally resistant to it. As it turned out, corn leaves contain an enzyme that detoxifies atrazine, and corn roots contain a substance that breaks it down.[2]

Less than three years after its discovery, and just a year after testing began in the United States, the U.S. Department of Agriculture gave Geigy the green light to tap the market it had been craving: the 65 million acres in the United States devoted to growing corn.

The company could not possibly have anticipated what a tremendous success the weed-killer would turn out to be. From 1959 to 1969, annual sales of atrazine and a less successful Geigy herbicide, simazine, by Geigy Chemical Corporation (the company's U.S. subsidiary), went from 15,890 pounds to 64.4 million pounds. Annual profits on the two herbicides soared from $1.9 million in 1961 to $90 million in 1969. From 1959 to 1969, in fact, Geigy Chemical Corporation racked up a staggering profit of more than $231 million on the two herbicides, even after paying its Swiss parent nearly $56 million in royalties. Geigy was already a giant in the industry when it merged with Ciba Ltd. in 1970 to form Ciba-Geigy.

Farmers love atrazine for the same reasons that make the chemical such a major threat to drinking-water supplies: It is long-lasting, and it does not dissolve in water. Farmers typically need to treat their crops with atrazine only once or twice each growing season and can spray in wet weather without having to worry that the herbicide will dissolve before it can kill weeds. Most important of all, atrazine is cheap. By 1969, after 10 years of sales, the price of atrazine had fallen from $4 to just $2.50 per pound. That still left plenty of room for profit, though, because per-pound manufacturing costs had fallen over the same period from 56 cents to 31 cents.[3] Today, atrazine is still far cheaper than its competitors.

Before the introduction of atrazine, only about a fifth of the U.S. corn crop was treated with herbicides. By 1988, the figure had risen to 96 percent, according to a survey by the Agriculture Department. Atrazine is now alone at the top of the heap: With U.S. sales of 68 to 73 million pounds in 1995, it is the best-selling pesticide of any kind in the nation.[4]

Atrazine is still crucial to Ciba-Geigy's financial health: It accounts for about 25 percent of the company's crop chemical business.[5] Although Ciba's patent protection for atrazine expired years ago, it still sells about 75 percent of the atrazine used by American farmers—and by a few homeowners, too, since atrazine is also available (but rarely used) as a lawn weed-killer.

Little wonder that Ciba-Geigy has fought so hard to keep atrazine on the market in the face of accumulating evidence about its dangers.

Since at least the 1970s, the company has known of studies that show atrazine causes cancer in laboratory rats. A study conducted for the company by Industrial Bio-Test Laboratories was "found by the registrant [Ciba-Geigy] to suggest an oncogenic response," as a scientist with the Environmental Protection Agency would later write, but the EPA discarded the results when IBT was found to be faking the results of hundreds of similar tests.[6] What apparently was the only earlier tumor study on rats ended inconclusively when almost all of the animals died of unrelated infections long before the end of the two-year period.[7] The test was not repeated, although a study conducted in the late 1960s by the National Cancer Institute concluded that atrazine was not carcinogenic in mice.

In 1984, more than a quarter-century after atrazine's introduction in the United States, the EPA finally took note of its carcinogenicity after a new study conducted for Ciba-Geigy showed that female rats fed atrazine were developing mammary tumors.[8] The same year, an Italian researcher discovered that women exposed to atrazine or other closely related herbicides were much likelier to get ovarian cancer than other women. He corroborated his finding with a larger follow-up study that found that exposed women developed tumors 2.7 times as frequently as unexposed women.[9] Then, in 1990, a Hungarian study showed that atrazine caused significant increases in the rate of benign breast tumors in male rats and of malignant uterine tumors, leukemia, and lymphoma in female rats.[10] Ciba-Geigy attacked the European studies as faulty; the company also maintains that the rat tumors are an aberration confined to a single strain of rat called the Sprague-Dawley.

Atrazine poses other dangers, too. It is so toxic that an adult male can die from ingesting two ounces of it. Some studies show that atrazine can damage DNA and induce gene mutations. (Ciba-Geigy disputes them.) Animals fed high doses of atrazine have suffered liver, heart, and kidney damage. Most alarming of all, recent studies have linked the chemical to hormonal changes—a finding that is chillingly consistent with the cancer studies, because the tumors have been found in organs such as the uterus and breast where hormone disruption can trigger unwanted effects, such as the rapid cell growth of cancer. While atrazine does not directly stimulate the growth of hormone-sensitive cells, as some other synthetic chemicals do, new research shows that by changing the way in which the body metabolizes estrogen, it may increase the risk of cancer. Indeed, out of 10 hormone-disrupting pesticides (including several known carcinogens) tested in a recent study, only DDT had as damaging an effect as atrazine on how the body metabolizes estrogen.[11] "Atrazine is not a direct estrogen, but it alters estrogen—and that's just as bad for you," says H. Leon Bradlow of the Strang-Cornell Cancer Research Laboratory, the study's author.

As the health threat was coming into focus, so was the evidence that millions of Americans ingest atrazine. It was showing up in supermarket corn—and in beef and milk, too, since cattle are routinely fed atrazine-treated corn.[12] More disturbing, there were increasing signs that atrazine had become one of the nation's leading contaminants of drinking water.

Ciba-Geigy knew as early as 1976 that atrazine was in the Mississippi River at levels of up to 17 parts per billion, especially during the early summer when rains washed the chemicals off newly sprayed fields.[13] Subsequent studies in the 1980s, first in Iowa and Ohio, and then later throughout the Midwest, proved that traces of the weed-killer routinely show up in rivers, lakes, aquifers, and even in raindrops.

In the Missouri River, for example, atrazine was found in 441 of 589 samples[14]; 165 were over the EPA's safety standard of three parts per billion. Atrazine was found in 448 of 580 samples collected in Illinois, at levels as high as 39 parts per billion—13 times the EPA's safety standard.[15] In a single 100-square-mile section of northeastern Iowa, atrazine was found in 96 percent of the groundwater samples and the vast majority of surface-water samples, too.[16]

The largest testing effort, by the U.S. Geological Survey, found atrazine in 990 of 1,604 water samples drawn from midwestern streams, rivers, reservoirs, and aquifers from 1989 to 1994. Even rainwater was tainted, the survey found. During the spring planting season, enough atrazine had vaporized off farm fields and into the atmosphere that the weed-killer could be detected in raindrops in 23 states and as far away as northern Maine.[17] In Iowa raindrops, atrazine concentrations sometimes exceeded 10 parts per billion, a separate study by the Iowa Department of Natural Resources found.[18]

The mounting evidence of atrazine's dangers has prompted countries around the world to ban it. Austria, Germany, Hungary, Italy, the Netherlands, Norway, and Sweden, among other nations, forbid the use of atrazine. But in the United States, which has always been atrazine's biggest market, the herbicide is as widely available as ever, even though the EPA has known for more than a decade about the cancer studies and the drinking-water-contamination problems.

The EPA did not classify atrazine as a restricted-use pesticide (which only farmers who pass a certification training course can apply) and limit its use near wells until 1990, four years after the agency had been informed about the cancer studies and two years after it declared the chemical a possible human carcinogen. In 1992, the EPA again ordered Ciba-Geigy and other manufacturers of atrazine to change the labels on their products, this time by lowering the dosages and restricting its use near creeks and lakes.[19] Home use of atrazine, a relatively small market, remained unrestricted.

Far from being aggressive moves by the EPA, however, the 1990 and 1992 changes were actually proposed by Ciba-Geigy, which was eager to head off tougher restrictions. The changes were minor indeed. Anyone, certified or not, could still walk into a garden store and buy atrazine. More than 90 percent of the nation's farmers already had taken the necessary certification course, and most were already getting the weed control they needed with less atrazine per acre than the maximum specified on the new label. The usage restrictions, meanwhile, applied only within 50 feet of wells and 66 feet of surface water.

Even with the lower dosages and other restrictions, the EPA estimates that midwestern corn farmers face a 1 in 863 lifetime risk of developing cancer from atrazine if they mix and apply it on their own, eat a typical diet, and get their drinking water from lakes, rivers, or other surface water (the way more than 50 million people in the corn belt get their drinking water). Nonfarmers who live in the Midwest face a 1 in 20,747 lifetime cancer risk, and a 1 in 13,850 risk if they also use atrazine on their lawns.[20] By way of comparison, the agency generally takes regulatory action when a chemical poses a lifetime cancer risk higher than one in a million.

The EPA took a tougher line with Ciba-Geigy in 1991 when the agency set a safety standard of three parts per billion for atrazine in drinking water. Many reservoirs and rivers in the Midwest routinely violate the safety standard during the spring growing season, when rains wash atrazine off fields and into water supplies. But water suppliers have been slow to install the costly filtration equipment that is needed to comply with the safety standard, which did not take full effect until 1996. It is easy to understand why: The potential price tag is staggering. Nationwide, the cost of complying with all of the EPA's standards for pesticides in drinking water could cost from $152 million to $465 million per year, with much of the cost attributable to atrazine, according to a study by the American Water Works Association.[21] Ciba-Geigy says that the cost is likely to be far less, but it has sued the EPA to try to force it to raise the safety standard from 3 to 20 parts per billion.

In November 1994, when the EPA finally initiated a special review of atrazine and two closely related herbicides, simazine and cyanazine, DuPont responded by agreeing to phase out cyanazine by the end of 1999. But Ciba-Geigy answered with barrels blazing. In March 1995, it submitted a 92-volume, 14,000-page response to the EPA; since 1983, in fact, Ciba-Geigy has spent more than $25 million and conducted more than 350 atrazine and simazine studies to bolter its case before the agency. The company has launched a massive campaign to pressure the EPA to keep atrazine and simazine on the market. So far, the agency has received more than 87,000 letters—the highest total ever about a pesticide.

The EPA is not expected to make a final decision until 1998 at the earliest.

Alachlor

The farm weed-killer market was already booming and crowded with products when Monsanto introduced alachlor in 1969. But within a decade, the white crystalline compound formally known as 2-chloro-2'-6'-diethyl-N-(methoxymethyl)-acetanilide was rivaling atrazine for supremacy in the world of herbicides.

In truth, alachlor is more of a partner than a rival to atrazine. Corn farmers embraced both herbicides because they killed grasses and broadleaf weeds, though alachlor tended to be more effective against grasses and atrazine more effective against weeds. Indeed, many farmers began using an alachlor-atrazine mixture, a one-two punch that made the two weed-killers the most popular in America. By 1989, Monsanto and Ciba-Geigy were selling more than 160 million pounds of alachlor and atrazine in the United States. It was an amount so vast that, by one estimate, alachlor and atrazine collectively accounted for about 15 percent of all pesticides used in the United States.[22]

As successful as atrazine was for Ciba-Geigy, at times alachlor was even more of a gold mine for Monsanto. Alachlor was the bigger seller in the mid-1980s (atrazine would pass it again in the 1990s), with 1986 sales estimated at a staggering $320 million.[23] Its biggest market by far was the United States, where Monsanto and other companies used it as the key ingredient in a host of macho-sounding products, including Lasso, Bronco, Bullet, Cannon, Freedom, and Lariat. (Another name, Alatox, was later abandoned, perhaps because it suggested—accurately—that the product was toxic.) Alachlor, in fact, was used on about 30 percent of the nation's corn and soybean fields, at rates of up to eight pounds per acre.

Monsanto's response was understandably frenzied when the EPA said in early 1985 that it was studying whether to ban alachlor. The company countered by submitting 20 bound volumes of documents to the agency as part of an all-out effort to save its immensely profitable product.

It would be no easy task. By the time the EPA initiated the special review in January 1985, the case against the chemical was strong. Studies in dogs showed that high doses of alachlor caused liver degeneration, and three separate high-dose studies in rats showed that alachlor caused kidney disease, eye lesions, and, in some cases, cataracts.

But the biggest threat was cancer. Two rat studies, one conducted by Monsanto and the other by a laboratory it had hired, concluded that alachlor caused nasal turbinate and stomach tumors in males and females, and thyroid follicular tumors in females. Brain tumors were also found in the rats, although it was debatable whether they were common enough to be statistically significant. In mice, another study found, alachlor caused lung tumors.[24]

A 1984 EPA report suggested, based on the animal studies, that alachlor applicators face a cancer risk of up to 1 in 1,000 (based on 30 days of exposure per year over a lifetime). A child who, over a lifetime, drinks a liter of water a day with just 1.5 parts per billion of alachlor runs a 1 in 100,000 cancer risk, the agency estimated.[25]

Meanwhile, evidence was beginning to accumulate that people in the nation's corn belt were drinking alachlor-tainted water. Even before the special review began, studies had found alachlor in water supplies in Nebraska, Ohio, and Ontario, and also in Iowa, where groundwater levels were as high as 16.6 parts per billion.[26] Many more studies were launched in 1985, and they quickly found alachlor in Illinois and other Midwestern states. The Illinois study, for instance, conducted from 1985 to 1988, found alachlor in 265 of 580 water samples at levels as high as 18 parts per billion, though the mean concentration was 0.24 parts per billion.[27]

By mid-1986, halfway through the EPA's special review, its Scientific Advisory Panel (a committee of scientists from outside the agency) had classified alachlor as a probable human carcinogen, and the EPA's Office of Drinking Water was pushing for an immediate ban.[28] Canada had banned alachlor in 1985, and Massachusetts was considering a similar ban.

Top officials of the EPA, under intense pressure from Monsanto, had other ideas. Relying largely on data collected by Monsanto, the EPA concluded that the actual cancer risk to applicators was close to one in a million—a thousand times less than the agency had estimated two years earlier. The agency also estimated that the lifetime cancer risk of drinking surface water containing the levels of alachlor typically found in the Midwest was about 1 in 250,000 but declared that there was not enough evidence to assess the health risks of alachlor in well water.[29]

When the EPA completed its review of alachlor in December 1987, it stopped far short of a ban. Instead, the agency reclassified alachlor as a restricted-use pesticide that only certified applicators may use. To reduce the risk of spills, the EPA also required that farmers who spray more than 300 acres per year with alachlor use sealed, premeasured canisters of the weed-killer instead of the resealable containers favored by growers who prefer measuring and pouring the chemical themselves. The EPA's new restrictions were not much of a burden, though: Ninety percent of all alachlor applicators had already taken the brief training course required for EPA certification.

The EPA still cannot say with any confidence how many drinking-water wells in the United States contain alachlor. Although its national pesticide survey has shown that alachlor may be present in as many as 101,000 rural water wells, the EPA has put its best-guess total at about 3,140.[30] Monsanto's own survey found that about 1,200 wells in the Midwest contain alachlor at levels higher than two parts per billion.[31]

Alachlor remains one of the nation's most popular weed-killers, although its market share is declining as Monsanto emphasizes newer

herbicides, such as glyphosate and acetochlor, that are more effective—
though more expensive—than alachlor and considered safer by the EPA.
The biggest blow to alachlor was struck not by the EPA but by Procter &
Gamble, the giant consumer-products company, which in 1991 declared
that because of the cancer concerns it would no longer buy alachlor-
treated peanuts for use in its peanut butter brands. Other major peanut
butter manufacturers followed suit, and peanut farmers quickly abandoned a product that at the time was being used on about half the nation's
peanut crop. But alachlor is still used by tens of thousands of corn and
soybean farmers and remains successful enough to be sold in more than
60 countries. Monsanto produces almost all of the alachlor used around
the world, even though the company's patent protection for alachlor
expired in December 1987.

Perchloroethylene

The process of removing soil and stains from clothing with chemicals is
known as "dry cleaning," but in truth the process is anything but dry.
The first alternatives to soap and water were, in fact, dry; the Romans,
for example, pressed absorbent powders and earths onto fabrics to
remove stains. But in the industrial age, the term "dry cleaning" came to
cover any washing mixtures other than water, which shrinks some fabrics
if the temperature is not controlled.[32]

The widely accepted story of the genesis of modern dry cleaning is
that a French tailor, J. Baptiste Jolly, discovered the cleaning power of
chemicals in 1825 when he accidentally spilled the paraffin from a lamp
onto a soiled tablecloth. He and his successors in the new dry-cleaning
(sometimes called "French-cleaning") business used turpentine, or a
distilled version, camphene, to remove soil and stains from garments.[33]

Whether they used paint thinners, fuel residues, or the chlorinated
hydrocarbon degreasers that are so widely employed today, dry cleaners
always have faced certain difficulties that arise quite naturally in the
business of using flammable, pungent, and caustic chemicals.

In Scotland in 1869, machines were developed to clean clothing with
the benzene that comes from coal tar—a residue of the most popular fuel
of its day and the first substance ever proved, through human disease
studies, to cause cancer. But the strong evidence that benzene also causes
cancer emerged decades after dry cleaners turned to other solvents.

As society shifted to oil and gasoline to fire the engines of progress, dry cleaners began using petroleum products. But these products tended, particularly in the drying cycle, to explode. Consequently, many cities passed fire ordinances to outlaw dry-cleaning establishments in residential areas. As a result, big cleaning plants moved to industrial areas. Corner shops became merely drop-off points, and customers would typically have to wait a week or more for their suits and dresses to be returned, smelling faintly like gasoline. During this period, according to Jerry Levine of the New York City–based Neighborhood Cleaners Association, the phrase "taken to the cleaners" gained popularity as an indictment of any business's poor treatment of customers.[34]

In the view of Levine and many others, the miracle chemicals of the World War II era allowed the dry-cleaning industry to reform its bad reputation. Perchloroethylene, introduced as a metals degreaser in Germany in the 1920s, could dissolve the dirt from a wool suit just as effectively as it removed grease from airplane parts. If the clothes were properly dried, they did not smell of perc. It did not expand or shrink fabrics, and it did not explode. Best of all, two huge chemical companies, Dow and Imperial Chemical Industries PLC, made plenty of it. Perchloroethylene is a by-product of manufacturing chlorofluorocarbons—CFCs were used in styrofoam, air-conditioning, and spray propellants until they were banned by international treaty in 1996 for eroding the earth's ozone layer—and, more recently, hydrochlorofluorocarbons.

Scientists knew in the early 1970s that perchloroethylene causes liver cancer in mice. On the basis of those tests, the Environmental Protection Agency classified the chemical as a carcinogen in high doses. However, the seriousness of the threat was not fully understood until the New York State Department of Health, acting on complaints from residents of buildings with dry-cleaning establishments, began taking measurements of perc fumes in apartment buildings. Its tests showed that perc levels in the apartments as high as 12 stories up far exceeded the limits meant to protect workers in the shops below.[35]

Later studies showed that if dry-cleaned clothing is returned home damp, perc fumes can permeate a closet, and simply "airing out" the closet will not help, because the gas evaporates only at hot temperatures. Yet other studies found high levels of perc in butter and other fatty items in grocery and convenience stores next to dry-cleaning establishments.

California, one of the few states to systematically study groundwater pollution, found that perc is even in tap water.[36] (Until the mid-1980s, cleaners in most states could get rid of their perc by pouring it down the drain; although the federal government has since made the practice illegal, some cleaners still do it to avoid the high cost of disposing of perc safely.)

Is such nonoccupational exposure to perchloroethylene dangerous?

A 1996 analysis by Consumers Union of the United States, the not-for-profit organization that publishes *Consumer Reports*, estimated that one of every 6,700 people who wear freshly dry-cleaned garments at least once a week could be expected to get cancer over their lifetimes from breathing fumes from the perchloroethylene left in the fabric. Not as dangerous as being married to a smoker, or living in a radon-contaminated house, either of which expose a person to a 1 in 500 chance of cancer, but many times more dangerous than the 1 in 100,000 risk posed by Alar, the apple pesticide that Uniroyal Chemical Company withdrew from the market in 1989, or the natural mold aflatoxin, which sometimes contaminates peanuts.

As for the risks posed by perc's penchant for accumulating in fatty foods, Judith Schreiber, a New York State health official, has predicted excess cancer risks as high as 1 in 5,000 for infants who are weaned on the breast milk of mothers who live in buildings that contain dry-cleaning establishments.[37] The few attempts to track perc's actual effects on humans show that these predictions, based on perc's effects on animals, are not off base. A study published by the National Institute of Occupational Safety and Health in 1994 found that rates of esophageal cancer among workers in dry-cleaning establishments are seven times the national average and rates of bladder cancer twice the national average.[38] And a 1994 study of people who drank perc-contaminated water in Massachusetts's Upper Cape showed five to eight times more leukemia than among their neighbors who had not drunk the tainted water.[39]

There are some 30,000 dry-cleaning establishments in the United States, and about 85 percent of them use perchloroethylene.[40] (The other 15 percent, mostly in Oklahoma, Texas, and other oil-producing states, still use petroleum cleaning fluids that create fire hazards as well as health risks.) The EPA notes that the dry-cleaning industry, which generates $6 billion a year in revenues, is one of the largest groups of chemical users to have regular contact with the public.[41]

More than 100 establishments in the United States offer a completely nontoxic alternative to chemical dry cleaning called "professional wet cleaning."[42] Computerization has made it possible to control temperature, moisture, and agitation to such a degree that cleaning fine garments with water—always possible in theory—has become practical.

But the dry-cleaning industry remains very much committed to perchloroethylene. For one thing, most shops already have major capital investments in the machinery needed to clean with perc (and which can be used *only* with perc.)[43] The EPA estimates that a third of all cleaners use old, first-generation equipment that requires workers to transfer clothes dripping with perc from a washer to a separate dryer.[44]

The three major producers of perc in the United States—Dow, PPG Industries, Inc., and Vulcan Materials Company—have been urging dry cleaners to switch to the newest machinery possible, arguing that it recycles the perc liquid and fumes, protecting both cleaners and the environment. In fact, it is largely because of this new equipment that the industry has cut its use of perc in half since the 1970s, even though the demand for dry-cleaning services has increased.[45]

But here's the downside: Through their purchase of the newer-generation equipment, many dry-cleaning establishments have locked themselves into the use of perc.

Formaldehyde

Most people know formaldehyde as the frog preservative they encountered in high school biology classes and as the embalming fluid favored by funeral homes. But of the billions of pounds of formaldehyde produced in the United States every year, most of it goes into the manufacture of wood products—chief among them particleboard, plywood, and veneer. Borden, Hoescht Celanese, Georgia-Pacific, Du-Pont, and at least a dozen other U.S. companies produce formaldehyde (at a very low cost) by heating methanol, a petroleum by-product.[46]

Minute amounts of formaldehyde can be found in the cells of all mammals; it is both produced and used up in the complex process of metabolism. As evidence has accumulated since the 1970s that formaldehyde makes people sick and possibly causes cancer, the chemical industry often has said that because it is a natural body product, low doses of it must be harmless. But formaldehyde has become such a popular chemical that everyone is exposed to it at levels far higher than the human body metabolized each day in the pre-industrial era.

In the 1800s, the scientific world was riveted by the discovery that researchers could re-create in the laboratory formaldehyde and other substances previously produced only by the human body, including the important chemical that would be formaldehyde's commercial partner, urea (the acid found in urine).[47] Mixing urea and formaldehyde, scientists created an amazing new material.

Urea and formaldehyde together formed a resin—a transparent, colorless, elastic, and insoluble goo—that scientists at first thought could be made into a glass substitute. Although urea-formaldehyde resin never proved stable enough for that, the glue was used to hold other things together. Mixed with sawdust, wood chips, and other wood scraps, it formed particleboard. Mixed with a soap, it turned into a foam that could be used as an insulation product because it hardened like styrofoam after application.[48]

The commercial use of urea-formaldehyde resins took off. Urea-formaldehyde foam was used, for example, to insulate the smoking room of the *Graf Zeppelin*.[49] And in the post–World War II housing boom, the use of particleboard and plywood helped to make homes affordable.

American workers, particularly those in the wood-products, textile, and metal-foundry industries, historically have had great exposures to formaldehyde. There is not adequate room in this book to describe the long battle over workplace regulation of formaldehyde, which involved many lawsuits by labor unions and intensive negotiations. The Occupational Safety and Health Administration agreed in 1993 to a tiered labeling system to warn workers that formaldehyde has been shown to be an animal carcinogen at high levels, an irritant at lower levels, and a possible irritant at still lower levels.

Throughout urea-formaldehyde's history, scientists have desperately tried to reduce what they often called the "odor problem"—the release of formaldehyde from the resin as a gas. Formaldehyde irritates, gags, and sickens anyone at high doses, and many people at low concentrations as well.

Scientists tried adding more urea to the mix. They experimented with porous folds of cellulose fiber to filter and hold the gas. They tried a two-step condensation process. No matter what they tried, they could only reduce—not eliminate—the smell. The resin always contains some formaldehyde that is not bonded tightly to the urea. Heat and humidity help to loosen the "free" formaldehyde molecules, and they escape from the wood as a gas—the same gas that biology students smell when they dissect preserved frogs. Over months and sometimes years, as the gas escapes from the wood, less free formaldehyde is left and the smell finally disappears.

However, the noxious smell was hardly the worst problem. Manufacturers had known since the appearance of reports in the medical literature of the 1920s that formaldehyde gas can be toxic—even lethal—yet it was not until the 1970s that the problem of formaldehyde in homes gained any public attention. The cost of heating homes and offices skyrocketed with the two Arab oil embargoes, and Americans began to seal and insulate buildings as never before. Tight homes and offices conserve energy, but they also allow toxic gases such as formaldehyde to accumulate inside. Some scientists believe the energy crisis also prompted manufacturers to spend less heat and time "curing" formaldehyde wood products, making them stronger emitters in the home, although those processes are trade secrets and that theory has never been confirmed.[50]

What is clear is that tens of thousands of families became ill in their homes, with flu-like symptoms, rashes, and neurological illnesses. At around the same time, new studies showed that formaldehyde had even worse effects: It caused rare nasal cancers in rats, for example.[51] By the late 1980s, the National Cancer Institute had found a heightened incidence of brain cancers in embalmers and anatomists and a quite rare malignancy, nasal-pharyngeal cancer, in industrial workers who regularly breathed high levels of formaldehyde gas.[52] Most of us, however, do not work with formaldehyde resins. We live with them—in our walls, furniture, flooring, and cabinets.

3

Science for Sale

"We are the data experts. We can handle the science."

—*Steve Spain, Ciba-Geigy Corporation's product manager for herbicides*[1]

DuPont "has shown a lack of respect for the civil justice system in general, which disrespect was evidenced by the attitude of DuPont's CEO, Mr. Edgar Woolard, that, and the Court paraphrases, 'When DuPont says what the science means, that is what the science is.'"

—*U.S. District Court Judge J. Robert Elliott of Georgia, in a 1995 ruling in which he fined DuPont $115 million for a "clear pattern of concealment and misrepresentation"*[2]

The rats gnawed at Clifford T. "Kip" Howlett.

"The issue is more than weak rats—it's dumb rats," Howlett wrote in an internal memo to his superiors at Georgia-Pacific Corporation in September 1980.[3] Howlett was in charge of safety and environmental affairs for the Atlanta-based forest-products giant.

Now, some rats threatened to make that a very difficult job indeed. Cancerous cells had begun to multiply in the tiny nasal passages of rats in a laboratory test chamber in Ohio. A second group of rats, bred carefully to be their genetic equals, twitched their tumor-free noses in an identical chamber nearby. The only difference between the two chambers was in

31

the air: The tumor-ridden rats were breathing formaldehyde six hours a day, five days a week.

Georgia-Pacific had a lot on the line. It was the nation's number-one manufacturer of particleboard and other low-cost wood products. If the Environmental Protection Agency judged the Ohio study a valid test of the human cancer risk from formaldehyde, it might require warnings or restrictions, or even ban formaldehyde outright, because the chemical was such a pervasive presence indoors, where Americans spend 90 percent of their lives. With so much riding on the outcome of the study, it is no wonder that Howlett was feeling the pressure and plotting an elaborate counterattack.

In the 1980 memo, Howlett was already laying out the essential elements of a strategy that he and others in the industry would pursue for years to come. They would argue that the rats were "weak"—that they developed the nasal tumors because of their poor physical condition, not because of formaldehyde. They would also argue that no one in his or her right mind would stay in a room with a formaldehyde concentration as high as that inflicted on the laboratory animals. Even the mice in the experiment, which had fewer tumors than the rats, knew enough to slow their breathing and "tuck their noses under their legs," as Howlett put it. The rats were "dumb," he said, because they maintained their normal breathing rates.

To bolster both points, the industry scientists would rely on a four-pronged plan. They would pay for a competing rat study that would be carefully designed to minimize the chance of another damaging finding. They would hire academic researchers to give "independent" testimonials to formaldehyde's safety and put the most favorable spin possible on test results. They would attack any scientist who said that formaldehyde was dangerous. And they would move aggressively to steer research in directions that would play down the chemical's risks.

In short, they would use the trappings of laboratories and the language of science in a decidedly unscientific venture: to stave off any real regulation of formaldehyde.

Ultimately, they would win.

The story of how science was used—and misused—in the formaldehyde controversy is no aberration. It is repeated over and over again in the long-running battles over the regulation of atrazine, alachlor, perchloroethylene, and a host of other toxic chemicals.

"It's easy to bias experiments," says Dr. David Rall, the retired director of the National Institute of Environmental Health Sciences, which is one of the few government agencies that tests chemicals for toxicity. "We live in a world of PR, and it's very hard to cut through it. NIEHS does some work, but most of the testing is done by the manufacturers, and there is this tendency to not test high enough doses and to not follow the test animals for a full two years, among other problems. We see it all the time, and it worries me."

The traditional definition of science as the acquisition of knowledge in the pursuit of truth just does not fit in the world of chemical regulation. In that world, science is a weapon. And no one employs it more effectively than the chemical industry, which has the deepest pockets and the most at stake.

Cheaters Prosper

The U.S. regulatory system for chemical products is tailor-made for fraud. The subjects are arcane, the results subjective, the regulators overmatched, and the real work conducted by—or for—the manufacturers themselves.

Those elements all came together in the scandal surrounding what was in the mid-1970s the nation's largest toxicology laboratory. Industrial Bio-Test Laboratories conducted 35 to 40 percent of all toxicology testing in the United States at the time, including the industry-financed tests that manufacturers submitted to the EPA and the FDA as evidence of the safety of thousands of consumer products, pesticides, and drugs.

The lab was exposed as a fraud in April 1976 when Adrian Gross, an alert FDA pathologist, started asking questions after he saw a rat study of the arthritis drug Naprosyn that looked too good to be true. In the ensuing months, federal regulators found evidence that dozens of its studies had been faked. Among other transgressions, sick animals were listed as healthy or were not included in order to achieve favorable test results. Some reports were total fabrications based on no study at all. Ultimately, hundreds of studies were declared invalid.

But until a team of federal investigators arrived at IBT's satellite lab in Decatur, Illinois, in January 1978 for a top-to-bottom inspection, there was no evidence that the fraud extended beyond the walls of IBT. Its employees may have thought they were doing what their clients wanted by faking results, but there was no evidence that the chemical companies actually knew what was going on.

That changed when the six-man team began interviewing employees and combing through whatever records had not already been shredded at IBT. In the report they wrote about the inspection, members of the team said they uncovered evidence that Monsanto, the manufacturer of alachlor and many other toxic chemicals, may have known about the fakery.[4]

"IBT is the worst anyone's ever seen," says Dowell Davis, an FDA pharmacologist who was part of the investigative team. "They were hell-bent on providing their clients with favorable reports. They did not care about good science. It was about money. They really had what was almost an assembly line for acceptable studies."

Paul Wright had been a research chemist at Monsanto before he went to work for IBT in 1971 as its chief rat toxicologist. Wright stayed at the lab for only 18 months before he returned to Monsanto with a new title: manager of toxicology in the company's department of medicine and environmental health. But that was long enough, the government investigators concluded, for him to be in the middle of a series of apparently fraudulent studies that benefited Monsanto products.

The investigators said that in one of the studies—for Machete, a rice and sugarcane herbicide—extra lab mice were added to skew the sample, a bit of trickery that was left out of the final report to the EPA. Even more serious accusations involved two rodent studies of monosodium cyanurate, an ingredient in a swimming-pool chlorinator, in which inspectors found evidence that missing raw data had been replaced by after-the-fact invented records, that animal deaths had been deliberately concealed, and that the final reports included claims about procedures and observations that never happened.

In all three cases, the investigators wrote in an internal memo, there was evidence that Monsanto executives knew that the studies were faked but sent them to the FDA and the EPA anyway. Wright's protocols for the three studies were riddled with mistakes. He had signed off on the completed studies after he returned to Monsanto as a supervising toxicologist.[5]

Wright was indicted, tried, and sentenced to six months in jail and two years' probation for his role in the IBT scandals, and he was fired by

Monsanto on conviction. The three tests were not part of the indictments. Instead, Wright was tried for fraud in connection with another Monsanto product: TCC, an anti-bacterial agent used in deodorants.

Monsanto executives vehemently deny that, before the cheating was exposed in 1976, anyone at the company other than Wright knew that the IBT studies were faulty. They point out that prosecutors could have brought charges against the company but did not. "Clearly, the government did not reach the conclusion that we had done that—otherwise they would have charged us," George Fuller, the director of product registration and regulatory affairs for Monsanto's agricultural division, says.

The IBT revelations were a major scandal, prompting front-page newspaper stories, congressional hearings, and soul-searching within the FDA and the EPA about their oversight of product testing. But it was not enough to stop the cheating.

In another celebrated case, Craven Laboratories, Inc., of Austin, Texas, was caught faking pesticide studies. To make sure that consumers are not ingesting unsafe levels of pesticides, the EPA requires manufacturers to measure pesticide residues on fruits, vegetables, and grains after they are harvested. Craven was a top residue-testing lab for Monsanto, DuPont, and other pesticide manufacturers.

This time, mindful of the IBT experience, the manufacturers themselves tipped off federal regulators after being told of the frauds by a lab employee. When EPA investigators first inspected Craven Laboratories in 1990, they could not find anything wrong because the company's manipulations of the data were so sophisticated. It was only after the whistleblowing employee showed the investigators precisely how the frauds were committed that they learned that Craven had faked studies of 20 pesticides used on more than 50 food crops. Fifteen employees, including the lab's owner, Don Allen Craven, ultimately pleaded guilty to fraud charges.

The chemical companies involved were not charged, and there is no evidence that they knowingly submitted false data to the EPA. There is evidence, however, that the companies were not as vigilant as they should have been in detecting Craven's cheating, according to the prosecutor in the case. "There were incidents that came out that indicated that the chemical companies might have, could have, ought to have found out about this earlier," says James Howard, who prosecuted the case for the

Justice Department and is now an assistant U.S. attorney in Maryland. "A lot of them had some egg on their faces and could have done a better job of picking up on this."

What about the EPA? It is supposed to police the nation's testing laboratories. But its oversight program is as minuscule as it is feeble. A 1991 investigation by the EPA's inspector general found that the agency had never inspected about 600 of the estimated 800 labs conducting pesticide-related studies in the United States and that it had audited only 1 percent of about 220,000 studies by those laboratories. Amazingly, many of the studies were not audited until after the pesticide was already registered, the report found. It also noted that since 1984, when the EPA's inspection program began, the agency had not assessed a single criminal or civil penalty for substandard laboratory practices.[6]

The inspection rate has gotten even worse since the issuance of the report. The number of labs doing pesticide-related work has swelled to 2,000, and about 1,550 of them have never been inspected by the EPA, according to agency records. The audit rate has improved slightly, to 3.5 percent, and the agency has fined 10 laboratories an average of $14,360. "There is no question we have not had adequate staff over the last few years to fully attack this problem," David Dull, the associate director of the agriculture and ecosystems division in the EPA's Office of Compliance, says. "There's no question that we don't have the resources to inspect all of these facilities in any reasonable time scale."

With the EPA providing so little oversight, it is no wonder that many apparent frauds are exposed by whistleblowers or by lawyers representing people who have been injured by toxic chemicals. In the case of DuPont's Benlate fungicide, for instance, C. Neal Pope, a Georgia lawyer, had been frustrated by the company's success in beating back lawsuits by hundreds of rose- and orchid-growing farmers—including four of his clients—who said that their crops were killed by small amounts of atrazine that had tainted supplies of Benlate. Unable to get enough evidence to prove the charge, Pope's clients had agreed to settle their case for $4.3 million instead of the $400 million they had been seeking.

When Pope learned that lawyers in Hawaii had uncovered evidence that DuPont had withheld or manipulated laboratory tests that suggested Benlate was indeed contaminated, he went back to the judge who had presided over his case. In August 1995, in Georgia, U.S. District Court

Judge J. Robert Elliott responded by issuing a blistering order that slapped DuPont with a $115 million fine. DuPont strongly denies the charges, and the company's appeal of Elliott's order was still unresolved in late 1996.

"Put in layperson's terms, DuPont cheated," Elliott wrote in his decision. "And it cheated consciously, deliberately, and with purpose."[7]

Spinning Science

Brazen cheating on the scale of IBT or Craven Laboratories is rare—or, at least, rarely uncovered. But subtle, sophisticated slanting of scientific research is part of the everyday strategy of chemical companies enmeshed in regulatory battles.

Lyle Jackson learned all about that in the summer of 1985. That is when he concluded that the so-called experts who had flown into Fayette County, Iowa, to test drinking-water wells for herbicide contamination were looking in the wrong places. The story of how that happened, a tale pieced together from internal EPA documents obtained through the Freedom of Information Act and from interviews with those involved, illustrates just how far a chemical manufacturer will go to make data come out its way.

Jackson had more than a passing interest in the issue. As the sanitarian, or chief public health official, of Fayette County, it was his job to ensure that the wells were safe. And as a longtime resident of the heavily agricultural county, he knew that dousing fields of corn and soybeans with atrazine, alachlor, and other weed-killers is a ubiquitous springtime ritual in Fayette County.

Like most Iowans, Jackson had never considered herbicides a threat to drinking water until 1984, when Richard Kelley had launched a study of contaminants in the state's drinking-water supplies. An official of what was known then as the Iowa Department of Water, Air, and Waste Management, Kelley had not planned to include atrazine and alachlor in his survey. But George Hallberg, a researcher at the Iowa Geological Survey, had found surprisingly high herbicide levels in water supplies in a small 1983 study, so Kelley decided to include the weed-killers in his statewide survey—and ended up finding them in almost every water source he tested. "When we got hits on alachlor and atrazine, the pesticide industry came down on us like a bunch of stormtroopers,"

Kelley recalls. "They came out and publicly criticized our study, and it was routine to get phone calls from their local reps telling me what an idiot I was."

Now, in the summer of 1985, Monsanto was spending $4 million to conduct its own nationwide monitoring studies under orders from the EPA, which was in the middle of its special review of alachlor. Kelley and Jackson did not like what they saw when they looked at the maps that showed which Iowa wells, lakes, and rivers Monsanto was sampling as part of its nationwide studies. "It was not really an objective type of study," Jackson recalls. In Fayette County, he said, Monsanto was sampling deep wells in clay soils where herbicides were unlikely to turn up, instead of shallow wells in sandy soils where they were common. Kelley looked at all the sampling sites in Iowa and came to the same conclusion. "The study was systematic—it was systematically designed not to find the product," he recalls.

Without speaking to each other about it, the men did something each had never done before: write the EPA. In his August 7, 1985, letter to the agency, Kelley noted that Monsanto claimed that the sampling sites "were picked in such a way as to maximize the potential for alachlor exposure."[8] In reality, he wrote, the opposite was true. Seven weeks later, Jackson wrote his letter. "Sandy areas of the county were carefully blocked out and avoided. Only those soil association areas with soils having moderate clay content and thick protective depths were included in the sampling area.... My intention is not to be negative about this study, but I believe it to be heavily biased based upon the selection procedure of the sample wells."[9]

Today, Monsanto contends that Kelley and Jackson were both wrong about the groundwater and surface-water studies. "Nobody's criticized the scientific validity of either of those studies," says Andrew Klein, a Monsanto research manager who helped to design the 1985 surveys. Lakes and rivers, he says, were chosen at random for testing at the EPA's request, while the nearly 250 well-testing sites nationwide were carefully selected to represent soil conditions across the United States. "We did not do a survey of Fayette County, we did a survey of alachlor use, and we looked at vulnerable and less vulnerable areas," Klein says.

The EPA's chief groundwater expert had a very different view of Monsanto's 1985 tests. In April of that year, Stuart Z. Cohen had written

Monsanto and listed in his letter the EPA's objections to the way the company had selected its proposed sampling sites. The project, he wrote, was not the "worst case" study Monsanto had promised. Because the study was being required by the EPA as part of its review of alachlor, the agency's "significant concerns... must be addressed before the study can begin," he wrote.[10] A month later, however, Monsanto responded with a letter declaring that the study "is well under way."[11] That was a surprise to Cohen and, apparently, to the rest of the EPA.

In a memo to himself that he inserted in the official EPA record, Cohen recounted his frustration in dealing with Monsanto's groundwater consultant, Olin Braids of Geraghty and Miller, Inc. "I also expressed my displeasure about the overall progress of the groundwater study," Cohen wrote. "This was triggered when Mr. Braids told me that *wells had already been sampled* [emphasis in original] in a half-dozen counties. I pointed out that we did not even have an agreed-upon protocol for the study.... I said that I could cover myself by stating in a memo to the file, which this is, that I disagree with the study, but my doing that won't help Steve Schatzow [the director of the Office of Pesticide Programs] make a decision next fall." Cohen added that "this is one of those instances where a chemical has fallen between the cracks of the registration standards and special review programs and where the necessary initiative to resolve the problem was not taken. I'm afraid it's too late to correct the matter this year, and we may have lost a whole summer's worth of monitoring data."[12]

"Stuart would go crazy over this stuff," remembers Arthur Perler, who, as the drinking-water office's chief of science and technology, worked closely with Cohen. "Everybody knew throughout the agency that to some extent the pesticide data was a joke because it was provided by industry. Monsanto was especially adept at throwing a lot of money at surveys. They clearly had a lot of data, and tried to convince us to use that data. They were not entirely forthcoming on how the data was collected, the quality of the data, and were helpful only in ways they wanted to be, and not in ways that the agency wanted at the time."

In their defense, Monsanto executives say that the company was facing an EPA deadline to launch the 1985 studies by the spring planting season. But in retrospect, they acknowledge, the company might have been better off asking the EPA for an extension to allow more time to work out an acceptable study instead of forging ahead despite Cohen's objections. "We're damned if we do and damned if we don't, and our best judgment was that it's better to go get some data than to sit here naked of data trying to defend this product," Patricia Kenworthy, the company's director of regulatory affairs, says.

By the end of 1986, Cohen had left the EPA, and Perler quit soon after that. Today, Cohen runs a small pesticide consulting firm; his major clients are golf course operators. He says that he only vaguely remembers the 1985 Monsanto dispute, which was resolved the following year, when Monsanto brought in a new consultant who devised a different protocol for a much larger $4 million groundwater survey that would not be completed until the early 1990s.

But just as Cohen had predicted in his memo, the questionable 1985 Monsanto studies were the most important groundwater surveys that the EPA had on hand in the fall of 1986 when it was time for Schatzow to decide whether to ban alachlor. Groundwater contamination was not the only issue that Schatzow needed to consider in deciding whether to ban the weed-killer, but it was one of the most important. Yet the EPA, stuck with the questionable 1985 Monsanto surveys, dodged it. Declaring that "the risks associated with alachlor exposure through groundwater cannot be adequately assessed at this time," the agency did not include tainted well water in its formal calculation of the health risks posed by alachlor, which the EPA decided not to ban.

Did Monsanto save its prize weed-killer from being banned by refusing to conduct the groundwater surveys the way the EPA wanted? There is no way to know for sure. Monsanto's Klein says that the EPA's decision not to ban alachlor had nothing to do with the controversy over the 1985 monitoring studies. Because the EPA did not think there was enough good data, he says, the agency did not calculate a specific estimate of the cancer risk from drinking water.

Even without that formal calculation, the drinking water issue was still part of the EPA's overall decision, according to Klein. "They did not say that the exposure from groundwater is x, but a very big part of their overall assessment was the potential exposure through drinking water," he says. "The nuance here is that they couldn't characterize it but they could assess it. They could evaluate it." Besides, he says, the EPA could have revisited the issue—but chose not to—after Monsanto completed an acceptable monitoring study in the early 1990s. That newer study concluded that about 1,200 wells in the Midwest contain alachlor at levels higher than the EPA's safety level of two parts per billion—a result very similar to the company's first study, he says.

However, Daniel Barolo, who now holds Schatzow's old job as the director of the EPA's Office of Pesticide Programs, says that the lack of good monitoring data may well have affected the agency's decision. There is no way to know whether the EPA would have taken a harder line with Monsanto if it had good monitoring data. But, he added, "anybody in a regulatory management position would tell you that it is difficult to get a handle on exposed populations at risk given the absence and the [low] quality of some groundwater monitoring studies."

And so Monsanto, by attacking independent monitoring studies and disregarding the agency's objections to its surveys, outmaneuvered critics such as Jackson, Kelley, and Cohen and muddied the waters at the EPA when it was time for the agency to decide what to do about alachlor.

Ten years before the alachlor controversy, formaldehyde manufacturers were following a similar path to ward off regulation.

Even before the release of the damaging 1980 study that had prompted him to complain about dumb rats, Georgia-Pacific's Howlett urged his superiors to put money into research because "the information that we assemble and the communications which result with Congress, agencies, and media will be the most effective way that we can influence the federal cancer policy."[13]

Once the 1980 study was completed, and formaldehyde had been linked with cancer, the need for action was urgent. The Ohio study, in fact, had been conducted by the new Chemical Industry Institute of Toxicology, which had given its own members, the manufacturers of formaldehyde, little oversight or control.

In 1978, executives of the big formaldehyde producers—Borden, Celanese, DuPont, and Georgia-Pacific—began to organize another group, the Formaldehyde Institute. For the next 15 years, this institute would be the place where the 40 to 50 companies that produced and used formaldehyde would hammer out their differences, then present a unified front in public relations, lobbying, and litigation. It would also be the mechanism through which the formaldehyde manufacturers could pool their resources and finance scientific research on their prized commodity, with much greater control over the studies than they had had with the Chemical Industry Institute of Toxicology. Their first task was to interpret the CIIT's rat cancer results in the most favorable light, much the way political strategists put a positive "spin" on events to make their candidates look good. The Formaldehyde Institute "will be putting together a blue-ribbon panel of scientists to review the final 24-month report" on the Ohio rat test, Howlett said in a memo to his superior, J. M. Nicholson. "That analysis may be too late."[14]

Actually, members of the Formaldehyde Institute had begun weighing the spin possibilities long before the CIIT had analyzed its final rat-cell slides. When the rats began showing cancer cells halfway through the two-year study, the chemical companies were required by law to report the findings to the EPA. In October 1979, the company then known as Celanese Corporation (which was later acquired by Hoechst AG, the German pharmaceutical-chemical giant, and renamed Hoechst Celanese Corporation) wrote a notification letter to the EPA, stressing that the results were "interim" and that the cancer-stricken rats had received the highest doses in the experiment.[15]

Despite the soft-pedal approach, other members of the Formaldehyde Institute criticized Celanese's filing, saying that it was "gratuitous and...not necessary" to refer to the CIIT's test as "having been conducted properly and that the results observed seemed to be valid," according to the minutes of an institute meeting in February 1980. That wording "should have been omitted from a legal standpoint," the minutes of the meeting said.[16] In other words, such statements would not help their lawyers as they prepared to attack the study before government regulators and possibly in the courts.

Alfred S. Cummin, the vice president of science and technology for Borden, Inc., decided to do some investigating of his own. His company, which is best known for making milk, cheese, and nontoxic children's glue (which originally was made by scalding milk), also was a major producer of glue for grownups: formaldehyde resin. Cummin contacted the breeder of the "Fisher 344" rats, the special experimental type used by the CIIT, and asked him about the virus that some of the rats had contracted halfway into the study.

The breeder acknowledged that no one knew how the virus would affect an inhalation study, Cummin reported to the Formaldehyde Institute at its March 1980 meeting.[17] By that time, Georgia-Pacific's Howlett already was on the "weak rats" beat. He had hired an expert to prepare a paper for the Formaldehyde Institute noting the possible impact of viral infection in the CIIT's study and emphasizing that the breed was known for spontaneous tumors.[18]

The Consumer Product Safety Commission, which at the time was considering a ban on formaldehyde insulation, delayed its decision for more than a year while it reviewed the Formaldehyde Institute's virus and spontaneous-tumor theories. Ultimately, the commission would judge them meritless. (Follow-up studies with other breeds in the 1980s by other researchers would duplicate the CIIT's original findings.)

Much more powerful and long-lasting was the Formaldehyde Institute's argument that some of the rats in the CIIT's study were exposed to unreasonably high doses of formaldehyde. Walter Benning, the president of the Manufactured Housing Institute, a member of the Formaldehyde Institute, dismissed the CIIT's cancer study in a February 14, 1980, story in the *Houston Evening Times*. "Breathing that in their test at 15 parts per million," he said, "is like breathing the stuff for 420 years in man."[19]

What the public never knew was that in a meeting a year earlier, officials of the National Cancer Institute had expressed concern to formaldehyde manufacturers that the level of 15 parts per million in the CIIT's study would be *too low*. They believed it was not the "maximum tolerated dose" level necessary to ensure that animal studies are statistically significant.[20]

Most Americans do not fully understand why scientists assault laboratory animals with such high doses of chemicals in cancer tests, and the chemical industry eagerly exploits the public's knowledge gap. But high-dose animal testing has been shown throughout the history of toxicology to be the only practical means of measuring the potential of a chemical to cause cancer, and it has been a highly reliable method as well.

Researchers cannot deliberately expose humans to suspected hazardous substances (though there have been such regrettable episodes throughout modern history). Instead, they test on animals. But this presents its own difficulties. A study conducted with a few hundred rats or mice costs nearly $2 million. And if the rodents are exposed to the low levels of carcinogens that humans face, a 200-to-300-rodent experiment would never be able to show the impact of even the most potent carcinogens, like arsenic, that produce one cancer in 10,000 persons exposed.

Scientists have determined, however, that exposing small numbers of rats to high doses of chemicals conveys well how low doses of those chemicals affect a large number of people. Scientists have tried other approaches over the years—testing chemicals on one-celled creatures or bacteria, for example—but none of these other methods predicts carcinogenicity in humans as accurately as do experiments on our fellow mammals.

The formaldehyde industry took all of these realities into account for the second rat test, which would be conducted by a New Jersey laboratory, Bio/Dynamics, hired by the Formaldehyde Institute. Just about the only aspect that would stay the same would be the species of rat—the same "weak" strain that the Formaldehyde Institute was criticizing elsewhere.[21]

Although cancer had not even begun to show up in the CIIT's study until 12 months into it, the entire Bio/Dynamics study would last only half that time. While the CIIT's highest dose of 15 parts per million was low enough to raise doubts among scientists at the National Cancer Institute, Bio/Dynamics would afflict its test animals with no more formaldehyde than three parts per million.

This level did not take into account the statistical problems in administering the dose in question to a small number of animals. And the Bio/Dynamics study certainly had a small number of animals: Compared with 120 rats and 120 mice in each group studied by the CIIT, the Bio/ Dynamics test used six monkeys, 40 rats, and 20 Syrian golden hamsters per level. The Bio/Dynamics study is an example of a phenomenon that Nicholas Ashford, a professor of technology and policy at the Massachusetts Institute of Technology, underscores in his frequent criticisms of industry's use of science. "It's possible to co-opt the system without telling a lie," he says. Using animals that are insensitive to cancer, keeping doses low, and shortening the duration of an experiment all can make chemicals seem less dangerous than they are.

Maintaining the low dose of three parts per million was apparently so important to the Formaldehyde Institute that it ordered Bio/Dynamics to scrap this "high-dose" group when levels crept up to four to six parts per million in 1979. On February 26, 1980, even though the reworked test had not been completed, John Clary, the director of toxicology for Celanese Corporation and the chairman of the Formaldehyde Institute's medical committee, announced the good news at a hearing of the Consumer Product Safety Commission in Hartford, Connecticut.

"A new study indicates there should be no chronic health effect from exposure to the level of formaldehyde normally encountered in the home," he said, according to a press release issued by the Formaldehyde Institute. "While a sensory irritant, [formaldehyde] does not cause any chronic or long-term health effects."[22] A few weeks later, James Ramey, the institute's chairman and director of product stewardship at Celanese, urged the rest of the industry to follow Clary's lead. "Full use of these results [must] be made with all regulatory agencies, and meetings with the [Department of Housing and Urban Development] and the Consumer Product Safety Commission are being scheduled," he said. "This is the first supportive information we have received, and it should be utilized to the fullest extent."[23]

The picture did not look quite so bright in July 1980 when Ramey told the Formaldehyde Institute's public relations specialists that there were certain health effects—acute sinus problems, as well as the cells that are a precursor to cancer—even at the relatively low "high" level of three parts per million.[24]

When the study was published in 1983 in a scientific journal, where it would be reviewed by scientists both inside and outside of industry, the full results were reported: the early signs of cancer cells in the rats, as well as decreased body and liver weights, beginning in the second week of the study.[25] But no press releases would explain these findings. The Formaldehyde Institute would continue to argue, and regulators would generally accept, that formaldehyde was not dangerous at low doses.

In early 1979, the Formaldehyde Institute had discussed conducting a joint study with the government, but only if it could obtain "some kind of [agreement] that the agencies need not proceed with independent studies."[26]

These discussions led to one of the most controversial formaldehyde studies of all: a five-year, $1 million joint research project by the National Cancer Institute and the Formaldehyde Institute. Regulators at the Environmental Protection Agency and the Occupational Safety and Health Administration frequently cite this research as evidence that formaldehyde does not cause lung cancer. The huge study of 26,000 factory workers concluded in 1986 did, indeed, find a 30 percent elevation in lung cancer deaths among those exposed to formaldehyde, but the scientists from the government, DuPont, and Monsanto would together conclude that this was no cause for worry. The rate of lung cancer did not increase with the level of exposure, and the excesses were not seen at all the industrial plants studied. "Factors other than formaldehyde might have been involved," the National Cancer Institute says in the fact sheet on the chemical that it distributes to the public.[27]

The fact sheet mentions that not all scientists agree with this conclusion, but it offers little hint of the depth of the doubts that were raised early on by the late Irving Selikoff, the director of the environmental science division of Mt. Sinai Medical Center in New York City. Selikoff, who was responsible for tracing the link between asbestos and lung disease in shipyard workers, expressed "serious misgivings" in a letter to the National Cancer Institute as early as 1980.[28]

Selikoff's primary concern was that the companies with an interest in the chemical would choose which workers would be studied under a confidentiality agreement, and that the National Cancer Institute's researchers would never receive raw data, such as the job history, the locations of the workers, information about the plants, processes, products and exposures, or evaluations of the death certificates. The institute's researchers had no way to check the accuracy of this data handed them by the manufacturers or to tell what judgment calls had been made in its production. No changes were made in the protocol in response to Selikoff's comments, however, and in 1986, his successor at Mt. Sinai, Philip Landrigan, would reiterate his mentor's concerns when the results were released. "The NCI's typically excellent analysis can in no way offset the possibility that there exist systematic biases in the data which were provided to them," he said in a letter that he sent to the Amalgamated Clothing and Textile Workers Union, which was concerned because the results were derailing its efforts to gain regulation of formaldehyde fumes in the workplace.[29]

Putting Up a Front

As skillful as they are in manipulating scientific data to justify the continued sale of toxic products, chemical companies have learned that spinning works best when somebody else is doing it for them. Manufacturers would always rather have someone else do the talking when it is time to defend their products.

The soothingly named American Crop Protection Association, which used to be known as the National Agricultural Chemicals Association, does most of the talking for the pesticide industry. Other industry-subsidized groups include the Center for Produce Quality and the Iowa-based Council for Agricultural Science and Technology. The Center for Indoor Air Research works for the tobacco industry, and the Risk Science Institute is financed by a variety of chemical companies. The International Fabricare Institute counts mom-and-pop dry cleaners among its members, but it also receives support from Dow, which manufactures perc.

Scientists from these research institutes testify before Congress, at government hearings, or in courtrooms, giving an appearance of detachment that their sponsors would never enjoy. The game has become so well understood on Capitol Hill that a few lawmakers have made a point of asking scientists—for the record—to state their affiliations. In 1994, for example, Representative Henry Waxman, a Democrat from California, brought to light that the work of one of the nation's most widely cited authorities on indoor air pollution, Gray Robertson, the president of Healthy Buildings International, Inc., was shaped by his company's largest client: the tobacco industry. That is why, more and more frequently, corporations are taking even more care to distance themselves from the researchers, while holding tight the reins of research.

Industry-financed institutes that stray from the fold sooner or later are herded back in.

The CIIT, which was founded in 1975 by 11 chemical manufacturers, initially took pride in its autonomy from the corporations that financed it. "The decisions on priority compounds [the first chemicals it would begin to test], and on the order and manner of testing, were made by CIIT quite independently of its member companies, just as CIIT plans the remainder of its research program," James Gibson, its director, wrote in *Formaldehyde Toxicity*, its 1983 compilation of research.[30] Although the CIIT regularly reported the progress of its formaldehyde study to members of the Formaldehyde Institute, the manufacturers expressed muted frustration at the arm's-length relationship. At a meeting in October 1979—at the time that Celanese was first reporting the rat cancers to the EPA—the Formaldehyde Institute's minutes noted that "CIIT will not allow Dr. Clary [of Celanese] to review the analytical work they have done and will, however, hire outside consultants who will be approved by the Formaldehyde Institute to do a review of their analytical material."[31]

The CIIT's formaldehyde–rat-cancer study would prove as durable as any in the field of toxicology. The International Agency for Research in Cancer cites it as the underpinning of its 1995 assessment of formaldehyde as an animal carcinogen and probable human carcinogen, and the EPA's Science Advisory Board has termed the study "unequivocal." Despite this success, however, the CIIT would never again conduct a two-year rodent study, which toxicologists regard as the classic test of carcinogenicity for any chemical. A remarkable amount of the CIIT's work since that time, in fact, has been devoted to research that casts doubt on the importance of its earlier finding that formaldehyde causes cancer in rats.[32]

Gibson gave an indication of this new direction as early as November 1981, when, in the New York City offices of Georgia-Pacific, he spoke about the study to the Formaldehyde Institute's public relations specialists. The PR group observed in its minutes that "Dr. Gibson remarked about the great chasm from animal carcinogenicity to human epidemiology," that his "clear, definitive" talk was "superb," and that he "will be contacted" to describe the findings to the media.[33]

A year later, CIIT researchers would be the key speakers at a scientific symposium cosponsored by the Formaldehyde Institute that, a trade publication reported, "largely exonerated" formaldehyde from responsibility for human illness.[34] Although symposiums generally bring people of various minds together to delve into a topic, the lecturers at the CIIT's gathering already were of one mind. One explained how mucous in the nose and throat acts as a natural shield against formaldehyde. Others described studies of DuPont factory workers and Canadian morticians that showed no connection between formaldehyde and human cancer. The Formaldehyde Institute's law firm, Cleary, Gottleib, Steen & Hamilton, forwarded the entire proceedings to the EPA as a "significant new epidemiological study" that, it said, showed there was little need for fear about human exposure to formaldehyde.[35]

"What happened in the 1980s is that the focus of CIIT's work changed, and formaldehyde may have driven it to some extent," Rory Conolly, the director of the CIIT's formaldehyde research program, says. "There was a strong interest in trying to understand the mechanisms causing the cancer."

As the CIIT shifted away from its early focus of evaluating the long-term toxic effects of the big commercial chemicals, its researchers began to do detailed research on how cancer happens, and the CIIT became known around the world as a hotbed of the "mechanistic" study.

On no single chemical would it do more of this work than on formaldehyde. By December 1995, scientists at the CIIT had published more than 125 scientific papers and 104 abstracts on formaldehyde. All this work led to the very same conclusion about "the great chasm" between humans and rats that Gibson had talked about 14 years earlier.

The difference was that now the CIIT had data on the complexities of how formaldehyde works on the DNA of rat and primate nasal tissue. Although the research is widely recognized as groundbreaking, there is some question over what it all means. The CIIT's science has, nonetheless, forestalled the federal government in the 1990s from regulating formaldehyde as a cancer-causing agent.

The EPA's use of the DNA data "was a milestone for the CIIT research program," Conolly and his colleagues wrote in a December 1995 report on their formaldehyde work. They explained the reasons for their numerous, many-faceted, and "iterative"—in other words, repetitive—studies of the chemical. When science does not know much about a chemical, government regulators tend to make conservative decisions to protect public health, they said. The 48 chemical manufacturers who pay for the CIIT's work have spoken with their wallets to the importance of the formaldehyde project. "Sustained funding has facilitated the development and maintenance of a critical mass of investigators and allowed the iterative examination of key issues," Conolly and his colleagues wrote.

The CIIT responds to the concerns of the companies that hold its purse strings, Conolly says, but not in a way that mars the integrity of its work. He says, for example, that the institute drew criticism from some in the chemical industry in the late 1980s for what seemed to be an exploration of the mechanisms of cancer for its own sake—in other words, without any practical purpose. Since then, he says, the CIIT has focused on scientific work that government agencies can use when making decisions on what to do about chemical risks.

But it is indisputable that the CIIT's $18 million annual budget is devoted to studies that seek to show that the risks are not so great for chemicals that already have shown signs of danger, even though there are at least 50,000 chemicals in commerce for which there is no toxicological data at all. And in contrast to the "independence" that Gibson boasted of years earlier, the chairman of the CIIT's board of directors, R. Hays Bell of Eastman Kodak Company, wrote in the institute's 1994 annual report of new efforts "to increase synergy" between its staff and member companies.

Buying Legitimacy

The nation's chemical industry, by pouring billions of dollars into the nation's research universities, exercises an overwhelming influence on the direction—and even the results—of chemical research in America.

Collaboration between industry and academia is nothing new, nor is it inherently troublesome. DuPont, for example, worked with university researchers during World War II—in a successful and beneficial partnership—to develop nylon.

But in those early years of the chemical revolution, when relatively few profitable synthetic compounds were on the market, DuPont and other companies focused their energies on developing new ones and bankrolled academic researchers who were doing the same. Today, so many immensely profitable chemicals have established niches in the marketplace that there is little incentive for companies to take a chance and support researchers who are looking for innovative products that might turn out to be far less profitable.

Instead, the industry pours its money into research that defends existing products—particularly those products that are dangerous to human health and the environment, says Nicholas Ashford, a professor of environmental health at the Massachusetts Institute of Technology. Instead of research on safer and cheaper alternatives, the academic community focuses on saving the market for chemicals in use today, because that is where the money is.[36]

The consequences are far-reaching. The chemical industry's influence is so pervasive that for any particular chemical there are invariably only a few, if any, independently financed studies to weigh against the juggernaut of corporate-backed science. As a result, even when regulators do view industry-sponsored research skeptically, there is little opportunity to get an alternative view. "The problem is there is no evidence on the public side," Ashford says. "Public-interest science has become underfunded, and the number of really independent, good academics is such a small number that industry is able to overwhelm the science. And industry's way of looking at the science is very unbalanced."

Just how unbalanced becomes clear through a review of recently published industry-sponsored studies of alachlor, atrazine, formaldehyde, and perchloroethylene.

At least 43 studies assessing their safety were financed by corporations or industry organizations (such as the Formaldehyde Institute) and published from 1989 to 1995 in major scientific journals in English, according to the National Library of Medicine's MEDLINE database. Six of them returned results unfavorable to the chemicals involved, and five had ambivalent findings. The other 32 all returned results favorable to the chemicals involved. In short, manufacturers batted .744 when they paid for the research.[37]

Their average would have been even higher except for some circumstances surrounding four of the studies about formaldehyde that reported unfavorable results. Three of them were bankrolled by the Health Effects Institute, which is financed not by formaldehyde manufacturers but by automobile manufacturers and the Environmental Protection Agency. The institute is studying the dangers of reformulated gasoline, which often emits formaldehyde. The fourth was financed by the Center for Indoor Air Research, founded by the tobacco industry, which seeks to show that chemicals are a more serious indoor air pollutant than cigarette smoke.

When nonindustry scientists did the research, the results were quite different. Governments, universities, medical and charitable organizations (the March of Dimes, for example), and other nonindustry groups sponsored 118 studies published on the four chemicals during the same six-year period. Except for the insurance fund of the Amalgamated Clothing and Textile Workers Union (the labor union now known as UNITE!), which sponsored two of the studies, these were organizations without a stake in any specific outcome. They included such federal agencies as the U.S. Department of Agriculture and the U.S. Air Force, state governments, foreign governments, and the United Nations. The results when government agencies and other nonindustry entities were paying for the research: About 60 percent of the studies (71) returned results unfavorable to the chemicals involved; the rest were divided between studies with favorable results (27) and results that were ambivalent or difficult to characterize (20).

How can the underwriters of studies have such an overpowering influence on their results? A close-up look at the arcane discipline of weed science suggests some important reasons.

Weed scientists—a close-knit fraternity of researchers in industry, academia, and government—like to call themselves "nozzleheads" or "spray and pray guys." As the nicknames suggest, their focus is actually much narrower than weeds. As many of its leading practitioners admit, weed science almost always means herbicide science, and herbicide science almost always means herbicide-justification science. Using their clout as the most important source of research dollars, chemical companies have skillfully wielded weed scientists to ward off the EPA, organic farmers, and others who want to wean American farmers away from their dependence on atrazine, alachlor, and other chemical weed-killers.

The numbers tell part of the story. About 1,400 weed scientists in the United States work for chemical companies, compared to just 75 for the federal government and 180 for universities around the nation, according to James Parochetti of the U.S. Department of Agriculture. Independent scientists "are a very small part of the picture, because herbicides dominate weed science so totally," says John Radin, who oversees weed science research at the USDA's Agricultural Research Service.

But industry's dominance of weed science is even more overwhelming than those numbers suggest. The reason is money. The USDA usually pays the salaries of researchers who work at the land grant universities that are at the heart of the nation's academic agricultural research. However, a scientist needs much more than a salary. Lab equipment, supplies, technicians, and graduate students are just as essential. Researchers must fend for themselves in securing the financial support for those extra expenses. If they are unable to win federal grants, there is generally only one other place to turn.

Chemical companies have the leverage to dominate research even with relatively small grants of a few thousand dollars per year. "Our universities are like a limousine with a well-trained chauffeur," Kent Crookston, who heads the department of agronomy at the University of Minnesota, says. "We have the limo and the chauffeur, but no gas money. When someone comes along with a little gas, they determine where we drive for a few thousand bucks."

"The large federal grants are difficult to obtain, and there's probably many more weed scientists out there who are relying very heavily on these smaller grants that come from the agricultural chemical industry," Alex G. Ogg Jr., a USDA researcher and past president of the Weed Science Society of America, says. "If you don't have any research other than what's coming from the ag chem companies, you're going to be doing research on agricultural chemicals. That's the hard, cold fact."

Some researchers have soured on the system. Take Orvin Burnside of the University of Minnesota, a past president of the Weed Science Society. A prominent weed scientist since the early 1960s, Burnside learned early that he could do little with the money his school provided: "If I was going to do something beside twiddle my thumbs after two weeks, I had to go out there and find some money." He found it in the chemical industry, which for years sponsored about 80 percent of his research, mostly on the herbicide 2,4-D.

But in 1993 Burnside had an epiphany. He wrote a scathing article for a weed science journal in which he called the discipline a "stepchild" of agricultural research and said that "public weed scientists must redirect their activities after four decades of largely herbicide-focused research."[38]

He also stopped accepting grants from the industry. "They like to direct what you do, and finally I decided I want to direct what I do," he says. "I'm going to work on what I feel is going to be most profitable and productive for the taxpayers, and that's not developing a new herbicide that the farmer's going to have to pay through the nose for." Burnside believes that herbicide use could be reduced by 50 to 75 percent without affecting crop yields. The nation's independent weed scientists, he says, ought to focus their energies on figuring out how to make that happen.

Conventional pesticide-based farming "is the only plate we offer farmers as a research community," Dennis Keeney, the director of Iowa State University's Leopold Center for Sustainable Agriculture, says. "The land grants themselves know that's wrong, and you read piece after piece by the land grant researchers that say they have to change. They talk, but they don't walk the talk. Not yet."

Few weed scientists have followed Burnside's example. Like most of his colleagues, Alex Martin, a professor of agronomy at the University of Nebraska, uses industry grants to pay for both his doctoral students and much of his research. "There is no question that availability of money influences research," he says. "You don't go to Monsanto to fund some kind of ecology problem. They're going to fund something in their area."

"The words 'conflict of interest' come to mind," the USDA's Radin says. "These weed scientists are honest people, but it's a very difficult position that they occupy when they don't have a very steady source of funding that allows them to be independent. It's an endemic problem."

Some forms of industry influence over academia cannot be measured concretely, such as the presence of chemical company executives on the advisory boards of the USDA research stations that are associated with major state universities. Others are theoretically measurable, except that no one is keeping track.

There is no way to know how much money chemical companies are funneling to universities and foundations for research, for example, but it is undoubtedly more than a billion dollars a year. The nation's land grant universities alone get more than $134 million a year in industry funds for agricultural research, with pesticides constituting a major share. In fact, just one of the thousands of potential sources of industry dollars, the corporate-financed Formaldehyde Institute, was spending at least $300,000 a year on research in the late 1970s and early 1980s—for just one chemical.

Even the chemical industry's charitable contributions—which, unlike its research grants, must be publicly reported—appear to be carefully targeted to serve corporate interests. Dow Chemical Company, for example, donated at least $24 million to universities and research foundations from 1979 to 1994, and the charitable arms of its smaller colleagues in the perc business—Occidental Petroleum Corporation, PPG Industries, Inc., and Vulcan Materials Company—gave a total of $7.5 million. During the same period, Monsanto Company's foundation gave at least $18.5 million, including $7.8 million to Washington and St. Louis Universities in the company's hometown. But most other beneficiaries of Monsanto's largesse were far-flung schools that specialize in agriculture and biotechnology—the fields where the company's business is concentrated. Formaldehyde users Hoechst Celanese and Weyerhaueser made research grants totaling more than $13.2 million. Borden's total was $2.8 million, while Georgia-Pacific gave $1.2 million and DuPont $500,000.[39]

The chemical industry also showers money on the nation's cancer research establishment. The American Cancer Society, the Arthur James Cancer Hospital & Research Center, the Ohio Cancer Foundation, the R.

David Thomas Cancer Research Foundation, and the Sloan Kettering Cancer Center all receive support from chemical manufacturers.

Industry officials say that their charitable giving is aimed not at influencing scientific research but at improving the quality of life in the communities in which their companies operate, although they acknowledge that there are public relations benefits as well. "Certainly the well-being of the communities where we operate and the company's image as a good citizen and a contributor to that community is very important," Monsanto's Kenworthy says. "It's very important. It's important to the welfare of our workers and their job satisfaction and how productive they are, and it's important as to how easy it is to operate in that community."

But it is difficult to ignore that, over the years, the nation's cancer research establishment has done much to bolster the industry's position that chemicals should not be blamed for disease. When the EPA was looking at formaldehyde in the early 1980s, Alan C. Davis, a lobbyist for the American Cancer Society, sent a letter to James Ramey, the chairman of the Formaldehyde Institute, in which he said that the society knew of no evidence that formaldehyde caused cancer in humans.[40] The Formaldehyde Institute presented this stamp of approval to the EPA.

In its December 1995 newsletter on research, the society criticized the Delaney Clause of the Food, Drug, and Cosmetics Act, which, until it was repealed in 1996, banned additives of carcinogens to processed food.[41] This zero-tolerance level, the newsletter said, ignores "the virtual absence of risk at very low-dose levels and the economic and nutritional benefits which can accompany the use of synthetic chemicals (preservatives, pesticides, sweeteners, etc.) in the production and processing of food." The newsletter article opened with the statement, "Risk is a natural part of life." It echoed the words of the Chemical Industry Institute of Toxicology, which in its 1994 annual report said, "Risk is a fact of life."[42]

Industry's friends in academia are similarly helpful. At the height of the controversy over a possible alachlor ban, in 1985, the Weed Science Society of America submitted a letter to the EPA that used some notably unscientific language to argue against any new restrictions. "Eliminating competition by regulation in the free-enterprise system is not the American way," it said. The society's letterhead listed eight board members who worked for universities or the USDA as well as its president at the time, J. D. Riggleman, who was then the president of DuPont's agricultural chemicals department.[43]

A decade later, the society leapt to the defense of atrazine when *Tap Water Blues*, a newly released report from the Environmental Working Group, a Washington-based environmental organization, was drawing nationwide attention to the problem of herbicide contamination of drinking-water supplies. The society issued a "position statement" that took a hard-line stance against stricter regulation of atrazine. It should not have been a surprise: A "senior research fellow" at Ciba-Geigy, Homer LeBaron, had been the society's president in 1989. LeBaron had written his doctoral thesis on atrazine and spent much of his 30-year career at Ciba-Geigy trying to expand use of the weed-killer until his 1991 retirement.

A friendly academic institution, the University of Missouri, even issued a press release in February 1995 that unblushingly began this way: "Atrazine, one of the Corn Belt's favorite chemicals, could get yanked from farmers' weed control arsenals by the Environmental Protection Agency because tiny amounts are getting into reservoirs and rivers used as public drinking water sources. Agricultural specialists have argued that the amounts found pose no health risks and that losing the chemical would double corn farmers' weed control costs." At a time when Ciba-Geigy was working feverishly to flood the EPA with testimonials for atrazine before the official comment period ended on March 23, the university even took the extraordinary step of including in its press release the EPA address to which public comments could be sent.[44] That same year, another university press release attacked *Tap Water Blues*, calling it "unfounded."[45]

The university's agricultural researchers may not have considered their pro-atrazine efforts particularly unusual. After all, they were used to working closely with agribusiness interests, including Ciba-Geigy. Over the previous five years, the University of Missouri's College of Agriculture, Food, and Natural Resources received more than $2.2 million in research funds from for-profit corporations, including more than $137,000 from Ciba-Geigy, university records show.[46]

The college's dean, Roger Mitchell, says that he worries about whether the school's farming research is too heavily focused on chemical use. "I accept that as appropriate criticism," he says. "I spend a lot of time thinking about it." He added, however, that the aggressively pro-atrazine press releases issued by the college were prompted not by a desire to repay Ciba-Geigy for its research grants, but by the college's view that Missouri farmers are reluctant to adopt non- or low-chemical farming techniques. "Those are very thoughtful choices that we make, and it isn't the chemical company that's driving that choice, it's our trying to evaluate what seems to be working for the farmer." Critics, of course, argue that farmers have been slow to embrace alternative farming techniques precisely because manufacturers and their allies in academia control the flow of information to growers.

A closer look at the work of the one of the most prolific researchers on herbicide pollution—David B. Baker of tiny Heidelberg College in Tiffin, Ohio—shows how industry funding can help to shape the scientific debate.

In the early 1980s, before chemical companies became a major source of Baker's funds, Baker's studies of surface-water contamination in northwestern Ohio were among the first to identify atrazine, alachlor, and other weed-killers as major pollutants. The EPA relied heavily on his work when the agency decided to begin its special review of alachlor in 1985. A review of Baker's papers from those early years shows that they tended to emphasize the seriousness of the herbicide problem. A typical sentiment was the first sentence he wrote in his summary for a paper he presented at a 1983 conference: "Relatively high concentrations of many currently used herbicides and insecticides are present in rivers draining agricultural watersheds."[47]

By the 1990s, Baker's tone had changed significantly. The herbicide contamination problem had become a major issue, and Baker was getting almost a third of his research money from Monsanto, Ciba-Geigy, DuPont, and other industry sources. In April 1994, for instance, in an "issue paper" he cowrote for the industry-financed Council for Agricultural Science and Technology, Baker concluded that the regulatory system "appears to be capable of maintaining overall risk at acceptably low levels" and that "there is no conclusive evidence that allowed concentrations are the source of any human health effects."[48]

Six months later, Baker was a key part of the industry's damage-control strategy after the Environmental Working Group's release of *Tap Water Blues*. The National Agricultural Chemical Association (later renamed the American Crop Protection Association) and other industry groups urged reporters writing about *Tap Water Blues* to call Baker, and many of them did so. A week after the report's release, the association flew Baker to Washington, D.C., where he delivered three speeches attacking *Tap Water Blues*—two to industry groups and the third to journalists at the National Press Club. Within four weeks he had written a 39-page rebuttal, and in the cover letter accompanying the critique he called the environmental group's report "erroneous and slanted" and "designed to frighten the public."[49] The pesticide association, as well as farm bureaus and manufacturers, also touted a paper Baker had not yet published in which he concluded that herbicides do not pose a significant health risk in water supplies. Less than a month after the release of *Tap Water Blues*, the EPA placed atrazine in special review, and Ciba-Geigy was facing an unprecedented threat to its 30-year-old gold mine. It was thrilled with Baker's study, published in early 1995, in which he and colleague R. P. Richards of Heidelberg, using unusually blunt language, wrote that "atrazine exposure through drinking water does not represent a significant human health threat."[50] In 1996, Baker was working on a sharply worded critique of another study by the Environmental Working Group of weed-killers in drinking water.

There is no factual contradiction between Baker's early and late work on herbicide pollution. What changed is his tone. His early government-financed research played up the extent of contamination; the later industry-financed studies played down the health threat posed by that contamination.

"There's certainly been some change in his tenor and approach," says George Hallberg, now the chief of environmental research at the University of Iowa Hygienic Laboratory and, like Baker, a pioneer investigator of herbicide contamination of drinking water. "One contrasts the rhetoric of the articles from early on to today and there certainly seems to be a difference."

Adds Richard Wiles of the Environmental Working Group, a frequent Baker antagonist: "He was one of the guys way back when who identified this as a problem. And the tone of his data, the interpretation of his data, has changed as his funding source has changed."

Baker denies that his findings have been influenced by industry grants to Heidelberg College's water quality laboratory, which he heads. The lab's 1994–95 budget projected the receipt of about $143,000 in industry funds out of a total budget of about $502,000, with the rest coming mostly from government grants.[51] However, while Baker says that he never allows funding considerations to influence the way he interprets scientific data, he acknowledges that funding can sometimes affect the way he presents that data. An industry-funded study, for instance, might stress the relatively low number of cancer cases linked to a chemical exposure, he says. "Do those [funding] concerns color the way I present the data? They may. But I try as hard as I can not to let those kinds of concerns influence the way I interpret the data." He also points out that scientists who rely on activist-oriented foundations for research support are vulnerable to the same criticism. "The funding issues work both ways with both kinds of groups," he says.

J. Alan Roberson, an engineer and director of regulatory affairs for the American Water Works Association, calls Baker an able scientist but says that his recent work grossly understates drinking-water con- tamination. "Industry spinning is more subtle than faking results," Roberson says. "These guys usually pull the samples and analyze the results straight up. The spin comes in the way they present the data."

An even more fundamental way that manufacturers spin academic research is by carefully selecting the scientists they finance. Like an investor picking a mutual fund, manufacturers tend to choose researchers who have a proven track record of publishing studies that reflect the industry's point of view. Ciba-Geigy, for example, turned to one of the chemical industry's most outspoken defenders in 1996 in an effort to counter new research by independent scientists who found that atrazine changes the way the body metabolizes estrogen in a way that increases cancer risk. Stephen Safe of Texas A&M University, whom Ciba-Geigy hired to conduct its hormonal studies of atrazine, has been one of the most prominent critics of the growing body of evidence that many chemicals dangerously mimic natural hormones or alter the way those hormones function in the body. Safe frequently debates the issue at scientific conferences and gatherings of journalists.

While corporations usually decide in private which scientists to finance, records of Formaldehyde Institute meetings provide a rare window into the process and show how carefully manufacturers shop for friendly researchers.

It is clear that before the institute wrote a $20,000 check to Dr. Leon Goldberg, who had just left his post as the CIIT's president, for example, it knew the gist of his forthcoming paper "Risk Assessment of Formalde- hyde—A Scientific Viewpoint."[52] The institute had already made plans to videotape Goldberg and to condense his paper for distribution to the news media.

Members of the Formaldehyde Institute also financed a pet project of Harry Demopoulos, then in the pathology department of New York University Medical Center. Demopoulos appeared on PBS's *MacNeil-Lehrer Report* to discount the first tests that linked formaldehyde to cancer. Demopoulos had asked the institute to provide $50,000 of the $220,000 in funds he needed for a symposium on "thresholds"—the theory that low doses of carcinogenic chemicals are safe.[53] On May 11, 1981, soon after word of the first rat cancer results on formaldehyde, Demopoulos wrote a letter to Ramey, the institute's chairman, in which he expressed outrage over this and other cancer scares. Demopoulos also noted that his fellow NYU researchers were working on a formaldehyde carcinogenicity study but that it was so poorly conducted it was unpublishable.

Arthur C. Upton, the chairman of NYU's Environmental Medicine Institute, was so disturbed by what he termed the "inaccuracies and implications" in Demopoulos's letter that he wrote a letter to the Consumer Product Safety Commission. Upton, the lead author of the paper that would, indeed, be published and back up the original evidence of formaldehyde's carcinogenicity, said that Demopoulos's statements were "not only unprofessional but intolerable in their implication that the work in question was incompetently performed."

When industry-financed studies are released to the public, there is often no disclosure of who paid for the work. For example, when formaldehyde manufacturers persuaded a federal court to throw out the Consumer Product Safety Commission's ban of their insulation products in 1983, the court cited the work of five scientific researchers who had found little evidence of formaldehyde-related health problems. The court pointed out that one of them, John Higginson, a scientist at the Universities Associated for Research and Education in Pathology, Inc., in Bethesda, Maryland, had formerly headed the prestigious International Agency for Research on Cancer in Lyon, France.[54] It did not, however, mention that two of the researchers, Gary Marsh of the University of Pittsburgh and Otto Wong of Tabershaw Occupational Medical Associates, a private firm, had worked for the Formaldehyde Institute and its member companies.[55]

Negative Campaigning

They are badly outnumbered and outspent by their industry counterparts, but independent scientists studying toxic chemicals can still have an impact when their work casts industry products in an unfavorable light. That is why manufacturers are not content just to generate their own favorable science; they have also launched elaborate attacks on researchers whose findings threaten their products.

Scientists who cross the industry often run a gauntlet of criticism. In congressional hearings, politicians grill them with questions helpfully supplied by industry lobbyists. In scientific meetings, industry scientists pepper them with hostile queries. If they submit their findings to a journal to be published, their work is often attacked in the prepublication review process by scientists employed by, or friendly to, the chemical industry. After publication, their studies are sometimes attacked in letters to the editor written by industry researchers.

Peter Breysse, a professor of environmental health at the University of Washington and an expert on human exposures to toxic chemicals, learned of this firsthand after he began working fervently to warn people of the dangers of formaldehyde in homes.

Breysse got interested in the topic in the late 1970s, after he received an unusual call about drugstore clerks who had fallen ill. On the way to the pharmacy, he wondered if the clerks had been sickened by a strong medicine spill. But when he arrived, the only thing he smelled was the unmistakable odor of formaldehyde—a smell that had been omnipresent, the clerks said, ever since the store had been remodeled. When Breysse peered at the store's shelves, he saw that they were wood but had no grain pattern. They were tightly pressed wood particles held together, he would learn, by formaldehyde.

Soon after, he was called to look into an illness in a mobile home. "Under the countertops, in the cabinets, there was a lot of formaldehyde," Breysse recalls. "It was from different manufacturers, but it was the same damn thing."

Breysse could not believe that so many people in homes and offices did not even know that they were exposed to this strong chemical. He began to study formaldehyde and its impact on health, and he spoke frequently on the problem. With every speech, he received more phone calls from people who had fallen ill. He visited homes and took more formaldehyde readings. Based on his research, he began recommending that the formaldehyde level in the air of a home be no higher than one-tenth of a part per million.

His work quickly attracted the attention of the Formaldehyde

Institute. "Arrangements were made to monitor the presentation being given on June 27 by Peter Breysse at the National Environmental Health Association," according to the minutes of a June 1979 institute meeting.[56] At an institute meeting four months later, Georgia-Pacific's Kip Howlett reported that two members, the National Particleboard Association and the Hardwood Plywood Manufacturers Association, had hired Irving Tabershaw, the director of one of the institute's favorite private laboratories, to meet with John Wilson, Breysse's superior at the University of Washington. "The purpose of the meeting . . . is to discuss the standards employed by Mr. Breysse in conducting his tests and publishing his findings concerning formaldehyde," the minutes say. "Mr. Breysse has in the past conducted purely clinical studies, merely measuring ambient formaldehyde levels after receiving a complaint."[57]

Tabershaw would criticize this method of research and offer the university a different approach to take in response to each formaldehyde complaint. The minutes do not describe this new "format," but they do make clear that Tabershaw would urge deeper study into whether formaldehyde really was the cause of the symptoms. That would involve allergists and other medical doctors, as well as toxicologists, diluting the emphasis on Breysse.

"It is hoped that this effort [Tabershaw's visit] will also give the University of Washington more than one expert in the field of formaldehyde," the minutes conclude. Breysse today says he feels that the administration at the University of Washington has always been supportive of his work, despite any criticisms they heard from industry representatives (or the fact that the school has received more than any other institution, more than $1.7 million, in charitable contributions from a particleboard manufacturer, Weyerhaueser.[58])

He does recall that school officials had agreed to meet with Formaldehyde Institute representatives, a meeting he also would have attended. "The idea was that we were going to meet with a number of other faculty members to discuss the possibility of doing some research in other fields—epidemiology, biostatistics," Breysse says, adding it was an effort he supported wholeheartedly. He knew that he had been criticized because his reports were only anecdotal—single case studies. Breysse had tried to do a systematic statistical survey of residents of mobile homes but could never get managers of parks to cooperate. Perhaps working with the industry would allow him or fellow researchers such access. But the

Formaldehyde Institute representatives never showed up for the meeting, Breysse says.

Scientific conferences are often the place where pesticide manufacturers launch their attacks against research that threatens their products, according to Shelia Hoar Zahm, a prominent pesticide researcher at the National Cancer Institute who has been the target of some of those industry attacks. "There's a lot of smokescreens, with people saying that this study is weak, or that one is weak," says Zahm, an environmental epidemiologist who has worked on a series of studies over the past decade that have linked the use of weed-killers—chiefly 2,4-D—to non-Hodgkin's lymphoma in midwestern farm families. "But if people think it through, the criticism does not invalidate the results. What I find all the time is that there is this raising of [a study's] limitations without discussing the implications of the results."

Criticism from manufacturers is virtually a reflex action, she says. At a 1992 conference in Saskatoon, Canada, for example, an industry scientist stood up and began attacking Zahm's new study of atrazine and non-Hodgkin's lymphoma—even though she had just told the assembled scientists that her research showed that atrazine was *not* linked to the disease. "He stood up and starting giving all the limitations of the study, saying the things they always say," Zahm recalls.

More often than not, however, the EPA does not see through the industry critiques. Ciba-Geigy's strenuous efforts to discredit the atrazine-related work of two European researchers, for example, have largely succeeded within the EPA.

During the 1980s an Italian scientist, A. Donna, had conducted a series of studies on women in the corn-growing Alessandria province of northern Italy, where at that time atrazine was the dominant weed-killer (it has since been banned there). In a 1984 paper, he reported that women who were exposed to herbicides were 4.4 times more likely to develop ovarian cancer. In a larger 1989 study, he found that herbicide-exposed women were 2.7 times more likely to get ovarian cancer. Donna also exposed Swiss mice to atrazine in a 1986 study and found significant increases in lymphomas.[59]

Ciba-Geigy attacked all three of Donna's studies. The company hired its own consultants to review his work and concluded that the studies were sloppily done and too small to be statistically meaningful. "The Donna paper has received a great deal of criticism among the scientific community," says Darrell Sumner, Ciba-Geigy Corporation's manager of health and safety issues.

By the time the EPA reviewed Donna's studies, Ciba-Geigy had assembled a thick dossier of criticism, and the agency ultimately decided not to include them in its assessment of atrazine's health risks. His work "did suggest an association, but there were some inadequacies in the study design, the carrying out of the study, and the researcher's interpretation," says Penelope Fenner-Crisp, the deputy director of the EPA's pesticides office.

The work of a second European, A. Pinter of Hungary, posed a far more serious financial threat to Ciba-Geigy. At a 1988 meeting of the International Agency for Research on Cancer in Lyon, France, Pinter stood up and reported that male rats dosed with atrazine got mammary tumors, females got uterine tumors, and both sexes developed leukemias and lymphomas.[60] In the audience that day was Reto Engler, who at the time was a senior science adviser in the EPA's pesticides office. He was impressed by Pinter's research, which had been funded by the IARC, a highly respected European-based agency whose studies are used by many countries around the world to set chemical safety standards. "IARC was very much in favor of this [Pinter] study, and when I came back to the United States I said, 'Hey guys, we should get that study,'" Engler, who is now a private consultant, recalls.

For Ciba-Geigy, the Pinter study was particularly damning because it was conducted in the Fischer 344 strain of rats. In its dealings with EPA, Ciba-Geigy had been asserting that atrazine causes mammary tumors in only one kind of rat, the Sprague-Dawley, because of a hormonal deficiency that is unique to the Sprague-Dawley strain. If the EPA accepted the Pinter study as legitimate, Ciba-Geigy's argument that the Sprague-Dawley tumors were a meaningless aberration would lose credibility. Acceptance of the Pinter study would also probably trigger an upgrading of the EPA's official estimate of atrazine's carcinogenicity. Atrazine was classified by the agency as a possible carcinogen because the chemical had been tied to only one kind of tumor (mammary) in one sex (female) of one strain (Sprague-Dawley) of one animal (the rat). But Pinter had found other kinds of tumors in both sexes of a different strain of rats, and under agency protocols that would likely lead to a reclassification of atrazine from possible to probable human carcinogen—and that reclassification would, in turn, make it much more likely the EPA would severely restrict atrazine or even ban it.

So the stakes were high indeed for Ciba-Geigy and the EPA when Pinter presented a summary of his results in France. But instead of contacting Pinter or IARC directly to get more information about the study, the EPA turned to Ciba-Geigy. "Ciba-Geigy, in fact, was instrumental in getting us whatever we could get for that study," Engler says. Based solely on the information provided by atrazine's manufacturer, Engler says, he and the other relevant EPA officials became convinced that Pinter's study was off base. A key issue, he says, was the fact that Ciba-Geigy told the EPA it was unable to get from Pinter the raw data upon which he based his conclusions.

Did the EPA ever try to contact the IARC or Pinter directly and ask them to respond to the criticism? "I don't think so, no," Engler says.

Says Ciba-Geigy's Sumner: "We tried to get the raw data, and that got caught up in the politics of the communist regime. . . . The Hungarian study was conducted during the demise of the communist regime, and they had some limitations in terms of their resources." He adds that when Ciba-Geigy did its own study with Fischer 344 rats, "we showed there were no tumors."

The IARC holds a very different view: It considers the Pinter and Donna studies valid. A special working group considered the atrazine issue in 1991 and concluded that those studies and others all fit together because they show that atrazine alters the way hormones function in people and animals.[61] Some regulators on this side of the Atlantic also agree. New York State, for instance, cited the Hungarian study in its decision to tentatively classify atrazine as a carcinogen. And when a puzzled Kenneth Bogdan, a researcher at New York State's Center for Environmental Health, wrote to the EPA to ask why it did not do the same thing, he got a curt reply. Henry Spencer of the EPA wrote that the agency had concluded that the IARC study was not even worthy of a written review because it "provided very little data that were usable to evaluate whether the data were completely or correctly reported."[62]

James Huff, a researcher at the National Institute of Environmental Health Sciences, has reviewed the IARC study and says he cannot understand why the EPA dismissed it without even a formal review. "I can't figure that out," he says. "This study can't be discounted simply because some of the protocols were not up to current standards. If you have a lot of cancer, you can't ignore the results."

Another expert who reviewed the Pinter study for several environmental groups agrees. "I don't see any problem with the study," says J. Routt Reigart, a professor of pediatrics at the Medical University of South Carolina, who is frequently called as an expert witness in lawsuits over chemical exposures. "It appears appropriately done. The statistics are appropriate, and their conclusions are appropriate. What it appears to look like to me is one of these bureaucratic processes that says that because you did not do your study the way we wanted, we're not only going to ignore it, we're not even going to further consider the issue."

Knowing that attack by industry is inevitable, many scientists have learned to attempt to predict criticisms and take them carefully into account when they begin research on a chemical.

Avima Ruder, an epidemiologist with the National Institute for Occupational Safety and Health, remembered what had happened to a 1987 study her agency had conducted into the health of 1,700 employees of dry-cleaning establishments whose names had been painstakingly gathered from union records in California, Illinois, Michigan, and New York. The results clearly showed that bladder cancer was triple the expected rate, but the EPA's Science Advisory Board had concluded in 1991 that this evidence "does not warrant designation of perc as a probable human carcinogen." The reason: Perchloroethylene manufacturers had countered with the argument that most of the workers had also been exposed to other chemicals, such as the petroleum solvents that dry cleaners largely abandoned in the 1960s because they tended to catch on fire.

In 1994, Ruder published a follow-up study on the health of the drycleaning group that she and her colleagues had so carefully assembled. Not only did bladder cancer remain high—2.54 times higher than expected—but deaths from esophageal cancer were double the expected rate. Even more significantly, she looked closely at the 615 workers exposed only to perc. She carefully sorted out those with five or more years of employment in the dry-cleaning industry, signaling significant perc exposure, and 20 or more years since they began working in dry cleaning, allowing a long enough time for cancer to develop. In this group, cancer of the esophagus was seven times higher than expected.[63]

Word of the findings began to circulate a year before they were published, after Ruder presented the results at a conference on women and occupational cancer in Baltimore on November 1, 1993. The study dominated the discussion at a meeting in New York City two days later that the EPA had called to gather information on dry cleaning's role in indoor air pollution and groundwater contamination.[64] The International Fabricare Institute and the Halogenated Solvents Industry Alliance, still quoting their critique of the 1987 study, were caught off guard.

In a letter to the EPA, the institute's top Washington lobbyist, Peter Robertson of the high-powered Washington law firm of Patton, Boggs & Blow (now Patton, Boggs), harshly criticized the agency for failing "to control the types of information" presented at the meeting.[65]

Robertson had called NIOSH's Ruder soon after the meeting and gathered information for an updated criticism of the new study. Ruder's efforts to isolate only those people who had significant perc exposure—a move that anticipated industry's criticisms—now became her study's biggest liability. She had "looked at small slices of the data . . . and found that in one slice of data there was an increase in esophageal cancer," Robertson said in a letter to the EPA. He pointed out that when all cancers were lumped together, the high death rates from bladder and throat cancers were offset by lower rates for other types of tumors. "Overall, there was still no statistically significant increase" in cancer, Robertson wrote.

Despite the sophistication of Ruder's second study, Robertson continued to argue that something other than formaldehyde might be to blame for the bladder and throat tumors. "Other causes could be present, such as drinking and smoking," he wrote, even though Ruder had already looked at those factors and concluded that they could not account for the high disease levels.

Meanwhile, Peter Voytek of the Halogenated Solvents Industry Alliance was telling the EPA that most of the dry-cleaning workers in the NIOSH study were minorities, who always have higher cancer levels than whites.[66] What Voytek did not mention is that more and more independent researchers think that minorities may be at greater risk than whites because they tend to live and work among more toxic chemicals.[67]

Voytek would use yet another tactic when a Massachusetts study also suggested that perc was a powerful human carcinogen.

Human studies are always vulnerable to attack because unlike the much-maligned rats, which are carefully bred and fed to minimize all outside influences except the chemical in question, human beings cannot be caged and scrutinized. But on the Upper Cape of Massachusetts, David Ozonoff, the chairman of the Boston University School of Public Health's environmental health program, studied residents who had been trapped in an unintentional experiment of unprecedented proportions. Since the 1960s, they had been drinking and bathing in water laced with perchloroethylene, an ingredient in a plastic compound that had been used to line their new water pipe to soften the harsh flavor of New England tap water.[68]

The study, led by Ann Aschengrau of Ozonoff's department, showed that the incidence of leukemia in residents with high perc exposure was five to eight times that in their neighbors who had not drunk from perc-lined pipes. Purely by accident, the people of the Upper Cape had participated in an unplanned, controlled experiment—a tragic misfortune for them, but a rare boon to scientists trying to understand perc's dangers.

Still, Voytek quickly moved in with criticism. He wondered whether the memories of the "next of kin" the Boston researchers interviewed for data on the dead were reliable. He pointed out that the number of cases was small, even though the percentages were high. He questioned whether Aschengrau had looked at enough residents. "It's fair to conclude that a kind of 'Texas sharpshooting'—shooting a hole and then drawing the bull's-eye around it—may have occurred," he says.

Ozonoff says that all of these possibilities were recognized and controlled so conservatively—as any good human study does—that any mistakes or miscalculations would tend to *underestimate* the disease linked to perc. That is why he believes that a true chemical health effect has to be very strong to show up in a human study. "A public health catastrophe is a health effect so powerful," he says, "that even an epidemiological study can detect it."

Steering Science

Asbestos doesn't hurt your health.
OK, it does hurt your health, but it doesn't cause cancer.
OK, asbestos can cause cancer, but not our kind of asbestos.
OK, our kind of asbestos can cause cancer, but not the kind this person got.

OK, our kind of asbestos can cause cancer, but not at the doses to which this person was exposed.

OK, asbestos does cause cancer, and at this dosage, but this person got his disease from something else—like smoking.

OK, he was exposed to our asbestos and it did cause his cancer, but we did not know about the danger when we exposed him.

OK, we knew about the danger when we exposed him, but the statute of limitations has run out.

OK, the statute of limitations hasn't run out, but if we're guilty we'll go out of business and everyone will be worse off.

OK, we'll agree to go out of business, but only if you let us keep part of our company intact, and only if you limit our liability for the harms we have caused.

> —*Boston University's David Ozonoff, a witness for plaintiffs in asbestos litigation, describing the series of defenses used by the asbestos industry*[69]

They can cheat. They can manipulate research results. They can create front groups, co-opt academic researchers, and attack independent scientists. Each kind of scientific manipulation by the chemical industry is, in its own way, effective in warding off regulators and keeping dangerous products on the market.

Using these techniques, chemical manufacturers set the fundamental direction of research and define the terms of the scientific debate. This is the most far-reaching effect of the chemical industry's misuse of science. Sometimes, the agenda-setting process is as simple as putting money in one place and not another. Pesticide manufacturers, for example, have been far more eager to spend money on developing new chemicals than on finding old ones in water and in soil. In fact, those chemicals had been in drinking water for decades before the scientific techniques to detect them were finally developed in the 1970s. Atrazine's most important metabolites (breakdown products), for example, are toxic and thought by some researchers to be as common in water as atrazine itself. But until the late 1980s, there was no way to know because Ciba-Geigy had not developed a way to test for them. Testing for a third important metabolite, hydroxyatrazine, did not begin until 1992.

Similarly, herbicide companies do not finance research on low-impact farming techniques that require fewer chemicals for higher crop yields. But they have been pouring money into genetic engineering research to develop corn, soybean, and other crop seeds that are designed to be used in tandem with chemical weed-killers. "It's definitely the market for the herbicides that is the driving force behind this, because the profit margins are much higher for chemicals than for seeds," Rebecca Goldburg, a senior scientist at the Environmental Defense Fund, says.

But the agenda-setting process goes beyond simply selecting research topics. It is also about setting the terms of the scientific debate. Industry, for example, has played a key role in helping to shift the focus of the federal government's toxicology research program away from the chemical-by-chemical tests that have proved so troublesome for manufacturers.

The federal government's National Toxicology Program has gathered toxicology data on only 10 to 20 percent of the 70,000 chemicals in commerce. And as science has advanced, its program of evaluating chemicals has slowed. In 1981, the NTP started two-year rodent studies for cancer-causing potential on 54 chemicals. In 1992, these expensive tests were begun on only eight chemicals. By 1994, the number had dropped to four.[70]

Congress stipulated in its fiscal 1994 budget for the National Institute of Environmental Health Sciences that the National Toxicology Program begin long-term testing on at least 15 chemicals.[71] The legislation underscored the importance of the institute's work, particularly in looking at public health issues that "would not be of interest to university-based researchers." But the National Toxicology Program's budget has not kept pace with the increase in the cost of research. Long-term rat studies cost $2 million apiece and take four to six years to complete.

Instead of looking at more chemicals, the NTP has chosen to look deeper and deeper into the chemicals it already has identified as hazardous. It is now conducting hundreds of studies on the mechanism of cancer. Kenneth Olden, the institute's director, has expressed his support for moving research into the mechanisms of cancer as a better use of limited resources. He organized a meeting on the topic in January 1995, in part as a search for "more accurate and inexpensive methods" for identifying hazards and assessing the level of danger.

At the meeting, Roger McClellon of the Chemical Industry Institute of Toxicology was one of the strongest voices in favor of the greater use of such research in regulatory decision making. "I'd like to underscore that the risk-characterization process really allows us to bring to bear the true richness of all the science, and all the shades of gray," he said. "We need to get away from the dichotomy...of where we talk about a carcinogen or a noncarcinogen. That simply denies us the opportunity to use the science we need to fully characterize risk."[72]

Ellen Silbergeld, a toxicologist at the University of Maryland who has worked for environmental groups, disagrees. "Much of our selection of chemicals on which we bring to bear mechanistic data has been less than scientific," she told the audience at the National Toxicology Program workshop. "We tend to deploy both public- and private-sector resources on mechanistic research into those chemicals that are positive in the rodent bioassays, and also, if I may say, those chemicals which are high-stakes."[73]

Even Olden's predecessor as the director of the National Institute of Environmental Health Sciences, David Rall, agrees. "Some of the mechanistic research is all right, but we've made it too easy [for manufacturers to argue that animal studies aren't relevant]," he says. "You can always develop a theory about why the animal data should be discounted, but you should have to prove it."

Among the mechanistic research that the National Toxicology Program has focused on in recent years: studies of benzene that show it to be more potent than previous rat studies showed, studies of benzopyrine that suggest the rat studies were right on target, studies of formaldehyde and diethylemene that show them to be less potent than rat studies showed, and studies of dioxin and steroidal hormones that are inconclusive.

Bryan Hardin, a senior scientist at the National Institute for Occupational Safety and Health, pointed to these results in sounding a warning against too much emphasis on this new wave of research. "The pursuit of more and better scientific data can be very effectively used by the interests and forces who are served by avoiding and delaying action," he says. "So long as we need more and better mechanistic data before we can make a decision, the status quo is protected, perhaps at the expense of the public health. Whose risk is being minimized? Once we get caught in this cycle—we need more information, we need better mechanistic data, we need better science before we can make a judgment—I believe it is very difficult to break loose. When, after all, is there ever enough information? I believe it may turn out to be like peeling the layers off an onion. We're not quite at the decision point yet. We're always another layer further. And we've known from the beginning that we had an onion."

The National Cancer Institute's Zahm agrees. "We should not get so hung up on the mechanisms," she says. "I mean, people still don't know what it is about cigarette smoking that causes cancer. The mechanistic data certainly helps, but it's not required to take action."

Chemical manufacturers, however; use mechanistic research to make the case that chemicals should not be considered cancer risks to humans. Consider the recent research on perchloroethylene and formaldehyde.

Scientist Trevor Green has researched perchloroethylene at Imperial Chemical Industries PLC's toxicology lab in Cheshire, England (which has been known as Zeneca Central Toxicology Lab since the company split in two in 1992). At an EPA public meeting on dry cleaning in New York City in 1993, a company official borrowed from the language of children's stories to explain this complex research. In the livers of mice, he said, perc breaks down and appears to generate "these little beasts" called peroxizones, which cause liver cancer. Because human beings do not generate "these little beasts," he went on to explain, "therefore, the mouse liver tumors in the perc study are not relevant to man."[74]

David Ozonoff, a researcher at Boston University who found an association between perc and leukemia, calls the "little beasts" theory "rank speculation." The International Agency for Research on Cancer was kinder with its assessment of the "little beasts" theory, calling it "plausible" but pointing out that perc and its cousin chemical trichloroethylene also seem to cause cancers outside the liver, where there are no peroxizones, and in ways that seem unrelated to the peroxizones inside the liver.[75]

The IARC concluded that the increased risks shown in studies like Ozonoff's "were unlikely to be due to chance," even though they were based on only a few cases. Taking all this evidence into account, the IARC upgraded its hazard rating for perc in February 1994—concluding that it was a "probable, not just a possible" human carcinogen.

On the advice of its Science Advisory Board, the EPA has treated perchloroethylene as "on a continuum between [possible] and [probable] carcinogen."[76] But the agency's staff was reconsidering its rating of perc in early 1996, trying to decide how it would treat the "little beasts" data.

The "how does it happen?" research into formaldehyde has made more headway at the EPA than have the studies on perc. All this research, which continues today, has been conducted by the Chemical Industry Institute of Toxicology in the years following its findings that formaldehyde caused cancer in rats.

Sara Schotland, a longtime attorney for the formaldehyde manufacturers, says that formaldehyde has been "a flagship chemical" on the question of whether risks to animals equal risks to humans. The CIIT's research, she says, has shown "you cannot assume that a high risk at high doses in rodents automatically translates to humans."

This is how the research works: Mercedes Casanova, a CIIT researcher, discovered that formaldehyde leaves a microscopic mark that shows where it has been, a bit of damage in the nasal tissue of a rat or monkey called a DNA-protein crosslink—DPX for short. She and her colleague, Henry D'A. Heck, developed a method of counting DPXs.

No one knows for sure what DPXs do. Two other CIIT researchers did test-tube studies that suggest that DPXs play a role in starting tumors. Counting DPXs was a good way of measuring exposure to formaldehyde, Casanova and Heck said. And although the rats had a lot of DPXs after breathing formaldehyde, monkeys had many, many fewer. As primates exhale, they eliminate formaldehyde before it can do damage to DPXs, Casanova and Heck argued. As with the "little beasts" in the perchloroethylene studies, their conclusion is that the rat's chemical cancer problems do not relate to those in man.[77] Members of the EPA's Science Advisory Board warned the agency in 1992 not to rely on mechanistic evidence as the basis for its decision making on formaldehyde.[78] The rats and monkeys in the CIIT's study were dosed with a single six-hour shot of formaldehyde. Humans, on the other hand, breathe formaldehyde in their homes, offices, and outdoor air every day. Moreover, most people do not breathe formaldehyde in a sterile test chamber but inhale it along with the tiny dust and pollen particles that are part of normal air. The studies of humans who have been exposed to formaldehyde amid these "particulates," to which the chemical has an irresistible attachment, show an association with lung cancer, the advisory board pointed out.

Some of the evidence that the EPA's Science Advisory Board cited comes from studies commissioned by manufacturers that have never been published in scientific journals and have never been made widely known.

The only hint of them is deep within the microfiche library at the EPA. By law, manufacturers are required to immediately notify the agency whenever they obtain information that a chemical could create a "substantial risk" to health or to the environment.

In May 1991, the EPA received a package of charts from American Cyanamid Company, which does not manufacture formaldehyde but uses it in its chemical production processes.[79] The charts showed that workers at the company's factory in Wallingord, Connecticut, who had been exposed on the job to both formaldehyde and particulate dust over the past 40 years had elevated rates of both the rare nasopharyngeal cancer and lung cancer.

The study had one big problem: Researchers did not know which workers smoked and which did not. It is a common problem because it is so difficult to trace the smoking histories of people long dead. But smoking causes so much of the lung cancer in any population that the scientific community thinks less of any cancer study that fails to take tobacco's deadly role into account. Consequently, smoking literally clouds any study of whether particular chemicals cause cancer. In an internal memo, the EPA scientists who reviewed the American Cyanamid study noted the possible confounding effects of smoking and said that more information was needed to draw more definitive conclusions. But the EPA received no further reports from American Cyanamid.[80]

There is a glaring weakness in this system of self-policing, in which the EPA relies on manufacturers to conduct and turn in "substantial risk" studies—the system that Congress developed when it wrote the Toxic Substances Control Act in 1976. Manufacturers face fines, or even criminal prosecution, if they fail to turn in data. But how does the federal government ever know if the companies are hiding something?

The EPA has grappled with this question over the years, and in 1991 it developed an amnesty program of sorts. It announced that, for a year, it would waive the big-money penalties for any manufacturers who turned in any old "substantial risk" studies that they should have provided to the agency earlier. In response, the EPA received 10,000 studies in 1991 and 1992 from U.S. manufacturers—studies that had never been published in scientific journals and that had never before been submitted to the agency.

Among at least 35 studies that dealt in some way with formaldehyde (some of these included formaldehyde but dealt primarily with other chemicals) was a 1985 study conducted in the Netherlands by DuPont that showed an association between cancer of the nasal sinuses and formaldehyde, especially in workers who also were exposed to wood dust.[81] This study, unlike the American Cyanamid study, had traced the smoking habits of the group and concluded that tobacco could not account for the cancer association.

DuPont included a legal memorandum with its submission stating that the company did not admit that the chemical posed a significant risk and arguing that it had not broken the law by not submitting the study to the EPA earlier. The memo charged that the agency had changed the rules in its amnesty program by calling for all "positive" findings—those confirming the existence of a hazard—without taking into account "the probability of its occurrence." DuPont, in other words, had judged the risks in the Netherlands study to be low.

The study itself, however, seemed to reach a different conclusion. Although the nose-throat cancer in the study, adenocarcinoma, is very rare in the general population, the study pointed out, it is "by no means a rare tumor type" for wood and furniture workers, who are exposed to high levels of formaldehyde and wood dust together. "There is a strong indication of the need for strict preventive measures in this occupational group," the study concluded.

At the same time that the DuPont and American Cyanamid studies were quietly turning up in the EPA's toxic-risks library, in 1991 and 1992, the agency was being plied quite regularly and aggressively with the studies from the Chemical Industry Institute of Toxicology that concluded formaldehyde was not that great a risk. And the CIIT continues its work on formaldehyde in an effort to respond to the many questions it has raised. Weren't the animals in the experiment exposed to formaldehyde far too briefly to tell whether the long-term effect on humans is the same? In 1995, Mercedes Casanova completed research that shows that the single six-hour shot of formaldehyde is enough to give results that would reflect chronic formaldehyde exposure.[82] The CIIT also completed a study of iron foundry workers in 1995 that concludes that any lung cancer deaths were linked to cigarette smoke and silica exposure, not formaldehyde.

Jerry Dunn, Hoechst Celanese's vice president for environmental, health, and safety affairs, says that the chemical industry's support of research into formaldehyde's effects was aimed at protecting public health. "We wanted to make certain that such a widely used chemical was not a hazard," he says. The results have convinced him and his colleagues in the industry, he says, that there is no cancer hazard for humans. The structure of the nasal passages in rats, unlike in humans, allows formaldehyde to concentrate. Human nasal passages provide enough protection, Dunn says, that low levels of formaldehyde can be quickly metabolized—in other words, used up—by the body, just the way the cells metabolize the formaldehyde the human body itself creates.

"Formaldehyde is uniquely part of one's body chemistry," Dunn says. "Without it, the blood chemistry would be upset." The only danger is receiving too much formaldehyde for the body to metabolize, Dunn argues, adding, "Nearly all chemicals, when you are exposed to an excessive amount, become dangerous."

Sara Schotland, an attorney for the formaldehyde industry, sums up the manufacturers' position this way. Nasal cancer—the cancer that appears in rodents exposed to formaldehyde—"is extraordinarily rare, and yet formaldehyde is ubiquitous," she says, arguing that if the chemical were dangerous, more people would have nasal cancer. The 22 published studies of formaldehyde's effects on humans, including embalmers, foundry workers, anatomists, and others, have been "exculpatory of formaldehyde as a human carcinogen," Schotland says.

What is the EPA's response to all this conflicting information? Positively indecisive.

The EPA has never formally adopted the Chemical Industry Institute of Toxicology's data on how cancer happens—research that suggests that formaldehyde is 50 times less potent than the agency originally determined in 1987. Unofficially, however, this view is widely accepted within the agency.[83] The EPA officials who developed the 50-times-less-potent assessment even had it published in a peer-reviewed journal, while noting that it might "underestimate" risk.[84]

Which study is correct—the iron foundry study or the American Cyanamid study or the long-lost DuPont study?

The International Agency for Research on Cancer considered all the evidence and concluded in 1995, just as it had in 1987, that formaldehyde is probably carcinogenic to humans.[85] The industry has urged the IARC to weigh mechanistic research more heavily in its hazard rankings, but the agency has held firm to its aim to evaluate evidence of carcinogenicity "independently of the underlying mechanisms," as it says in the preamble to each of its monographs.

The preamble also explains why the IARC relies so heavily on evidence of carcinogenicity in animal tests. "All known human carcinogens that have been studied adequately in experimental animals have produced positive results in one or more animal species," it says. "Although this association cannot establish that all agents and mixtures that cause cancer in experimental animals also cause cancer in humans, nevertheless, in the absence of adequate data on humans, it is biologically plausible and prudent to regard agents and mixtures for which there is sufficient evidence of carcinogenicity in animals as if they presented a carcinogenic risk to humans."

The long fight over the carcinogenicity of atrazine is a case study in how the industry exercises its power to set the scientific agenda. Whenever new information has surfaced about atrazine's dangers, Ciba-Geigy has been able to shift the debate in a process that resembles nothing so much as a jittery suspect being questioned by an inept police detective: When one alibi falls flat, the company simply offers a different one, and each time the government spends years laboriously checking each story out.

For many years, the manufacturers of atrazine had a simple position: Atrazine did not cause cancer—at any dose, in any kind of animal, under any circumstances. It was a hard position to contest, since there was virtually no publicly available data on the issue. The only early rat study of atrazine's cancer-causing potential, conducted in 1961—two years after atrazine's introduction in the United States—by a consultant to Geigy, was a failure: Almost all the rats died, apparently of unrelated infections, long before the end of the two-year period.[86] The only other test, conducted by the National Cancer Institute in 1969, concluded that atrazine did not cause tumors in mice.

By the mid-1970s, the situation had finally changed, but Ciba-Geigy did not have to change its company line that atrazine does not cause cancer. Under the new Federal Insecticide, Fungicide, and Rodenticide Act, the company had to conduct a cancer study and many other tests and submit the results to the EPA, which would decide whether to allow atrazine to stay on the market. This time, the new cancer study suggested "an oncogenic [tumor-causing] response," as an EPA scientist would later state in a vaguely worded memo. But the company was able to keep arguing to the contrary because, as it turned out, the test had been conducted by scandal-ridden Industrial Bio-Test Laboratories.[87]

That an industry-financed study would find cancer was particularly noteworthy. "If IBT finds a positive carcinogen, you can believe that more than if IBT finds a negative one," Reto Engler, the former senior science adviser in the EPA's pesticides office, says. "It's easier to sweep negative studies under the carpet, but if it's positive it hits you in the face. If an IBT study was positive, you could bet your child's head that the chemical was positive."

When the EPA finally got around to reviewing the atrazine study in 1982, it concluded that Ciba-Geigy would have to redo the study. That was another break for Ciba-Geigy, because it bought the company more time. "If we had had a cancer study of sufficient integrity at that time, we might well have started the risk assessment process sooner," says the EPA's Fenner-Crisp, currently the deputy director of the pesticides office.

Ciba-Geigy finally had to shift its position in 1984, when, as required by law, it told the EPA that the new two-year study was halfway along and that high numbers of mammary tumors were appearing in female rats—the same kind of tumors that appeared in the discarded Industrial Bio-Test study conducted almost ten years earlier. Far from settling the question, however, the new study launched the company on a two-pronged scientific quest that still continues today: to prove that the breast tumors are a fluke confined to a single strain of rat, the Sprague-Dawley, and that low levels of atrazine pose no cancer risk because a threshold dose is needed to cause tumors.

Monsanto has embarked on a similar quest for alachlor and has come up with three elaborate explanations about why each of the three kinds of cancer that alachlor causes in rats—nasal, thyroid, and stomach tumors—would not occur in humans exposed to the chemical. The EPA still has not formally changed its view that alachlor is a probable human carcinogen, but in 1996 it adopted a new way of calculating alachlor's cancer risk that has the effect of playing down that risk.[88]

As with alachlor, formaldehyde, and perc, the EPA has allowed the atrazine debate to take a detour to arcane questions of the mechanisms of carcinogenicity because conventional cancer studies did not go the company's way. Ciba-Geigy has spent millions of dollars on a series of studies aimed at showing that atrazine is safe because, the company argues, it is carcinogenic only above a threshold dose that is never exceeded in lakes, rivers, and aquifers tainted with the herbicide. It has also fed atrazine to mice and a different type of rat, concluding each time that the chemical does not cause tumors. So far, the EPA still says that there is not enough evidence that a threshold exists for atrazine. That is not surprising, because the notion that toxic chemicals require a certain dose before they "switch on" and become carcinogens is extremely controversial. The only compound that the federal government has ever regulated based on a carcinogenic threshold, in fact, is saccharin. "At this point in time, they [Ciba-Geigy] have not yet convinced us their threshold hypothesis is correct, although quite frankly we have always thought it has merit," the EPA's Fenner-Crisp says. "We've been quite open about that."

Even so, Ciba-Geigy has succeeded in creating enough uncertainty to forestall regulation, despite the long-standing evidence that the herbicide poses a cancer risk. Thirty-six years after atrazine went on the market and began leaching into the nation's water supplies, the EPA still says that the question of whether it causes cancer in humans is unresolved.

4

Keeping the Watchdogs on a Short Leash

Along the banks of the Potomac River in January 1993, federal environmental regulators rounded yet another turn in what had become, seemingly, a journey to nowhere. Packed into a meeting room in Alexandria, Virginia, were all the industry interest groups and all the tension, generated by a dozen years of government efforts to deal with the possible hazards of one of the most common chemicals in the American home.

Like so many of the debates that had slowed the course of deciding what to do about formaldehyde, this issue was subject to endless argument and at times was even faintly ridiculous. The government simply wanted to measure formaldehyde gas in the "typical" home. But industry officials had steered them into a rumination on one of life's eternal questions: Do yuppies cook?

If so, some technicians and lobbyists at the meeting said, formaldehyde levels in the home could be elevated, because the higher the temperature, the greater the leaching of formaldehyde from kitchen cabinets and other wood products. Particularly if these prototypical families fry their food, they argued, it would be difficult to judge how much of this irritating, possibly carcinogenic gas comes from the floors, walls, and cabinets.

"That generation spends less time in the kitchen than any other community, and that is a fact, and that they don't open the doors to cabinets a whole lot," C. Richard Titus, the executive vice president of the Kitchen Cabinet Manufacturers Association, said. He was urging the EPA not to bother monitoring kitchens for formaldehyde gas. William McCredie, then the executive vice president of the National Particleboard Association, summarily dismissed this sociological analysis. "I have data that refutes that," he said. His manufacturers were angry at the EPA for focusing on walls and flooring to the exclusion of kitchen cabinets.[1]

The afternoon wore on with ever more esoteric debates over the potential difficulties of measuring the mix of chemicals that regrettably had become standard in the air of the American home. And the EPA got no nearer to its goal of deciding once and for all what to do about formaldehyde.

Ever since the federal government began addressing environmental hazards in the early 1970s, industries have sought to portray regulators as power-crazed bureaucrats whose senseless edicts are sucking the lifeblood out of the American economy. The facts, however, suggest otherwise.

By 1988, for example, the Occupational Safety and Health Administration had managed in its 18-year existence to set limits on only 24 of the thousands of chemicals in the American workplace, and only after lengthy, stop-and-go lawsuits on each. In 1989, when OSHA tried to set standards for 200 chemicals at once based on well-established research, the courts again struck down the agency's directive. By 1992, the EPA had completed health reviews on only two of the more than 19,000 pesticides that already were on the market before tougher testing standards went into effect in 1973—reviews that by law it was supposed to have completed by 1976. The General Accounting Office noted in a 1992 report, "At the current pace, not until early in the next century will the federal government be able to provide assurance to the public that these pesticides are indeed safe to use."[2]

In many cases, the government does not even try to act anymore without the blessing of its potential opponents. Meetings like the formaldehyde parley, with improbably large arrays of overlapping and competing interest groups, have become the norm. And in trying to please everybody, federal regulators seldom please anyone.

Ironically, industry does not much like the process that it has helped to create. Every major chemical company spends hundreds of thousands of dollars a year on regulatory lawyers to fence with the agencies, or to sit in on meetings as what the Bush administration referred to as "stakeholders" and what the Clinton administration sometimes has called "industry allies."[3]

If the "public" is represented at these meetings at all, it is not by the EPA officials who are on the public payroll. Instead, the EPA acts as a mediator and brings in representatives of one or two environmental organizations, or perhaps a researcher who has taken an interest in a particular hazard, to represent the interests of all Americans. The industry does not like dealing with these appointed public representatives, who often are veterans of battles with corporations in the courtroom or in the lobbying halls of Congress.

And the debates never end.

Members of the National Particleboard Association complain that after they worked out a voluntary standard on formaldehyde fumes from their products with the Consumer Product Safety Commission in the 1980s, the EPA called them together in 1993 to talk about testing their products for the same thing.

Perchloroethylene producers worked with the EPA's Office of Air and Radiation to come up with rules to control pollution from dry cleaners. Then another EPA department, the Office of Pollution Prevention and Toxics, stepped up its work on the same issue, which it viewed as a more serious problem, forcing the industry to argue its case once again.

And throughout the long regulatory battles over atrazine and alachlor, manufacturers have been pulled in three directions by the conflicting standards and priorities of the EPA's pesticides and drinking-water offices and the U.S. Department of Agriculture.

It is difficult to argue with industry officials who say that time and money are wasted in this process. Still, a closer look at the tortured histories of government efforts to protect the public from these chemicals shows that, wittingly or not, this is precisely the system that the big corporations have helped to create.

Manufacturers have become addicted to a system fraught with delay, because delay has served them so well. With large budgets and long tenures, corporate executives can wait out and wear down regulatory agencies, where rapid turnover at the top assures that no high-ranking decision maker stays on a single issue for too long.

Wearing Down the Regulators

The walls were closing in on Monsanto in early 1984. The EPA was moving aggressively against the company's top pesticide, alachlor, which had just been definitively unmasked as a cancer-causer in lab animals and would soon be officially classified as a probable human carcinogen.

At a May 24 meeting in Washington, EPA regulators huddled with their Canadian counterparts "to make sure both countries are headed in the same direction," according to an internal agency memo summarizing the meeting that was obtained through the Freedom of Information Act. The EPA told the Canadians that in mid-July it would issue a notice "proposing cancellation of all alachlor registrations"—in other words, a complete ban on what was at the time one of America's best-selling pesticides—and the Canadian regulators told the EPA that they were planning to do the same.[4]

Just a few months earlier, the EPA had even considered an emergency ban that would have taken effect immediately, but it decided to wait until summer to avoid disrupting the spring spraying season. "There were a lot of folks on our team who wanted to cancel that sucker," remembers Paul Lapsley of the EPA, who attended the meeting with the Canadians and is now the agency's director of regulatory management. In that role he supervises the process under which the EPA establishes 300 to 500 new environmental rules per year.

When a trade newsletter reported that the participants in the closed-door meeting were ready to ban alachlor, a Wall Street analyst downgraded his rating of Monsanto's stock, and the company's share price fell so quickly that, on June 7, the New York Stock Exchange briefly suspended trading of Monsanto stock. "Whenever we had a news leak like that it was always a mess," Lapsley says. "You get calls from the White House and Congress and everybody's going crazy. They realize something's going to happen, and everybody wants to know what it is you're going to do."

Amid the tumult, Monsanto arranged a quick meeting with EPA officials and pressed them to slow down their review. The officials assured Monsanto that, despite what they had told the Canadians, they still had not decided what to do about alachlor.[5] By then, Monsanto was conceding that alachlor was an animal carcinogen. But it was promising better training programs for alachlor applicators and was pouring money into studies designed to show, among other things, that the alternatives to alachlor were less effective and just as dangerous, that the costs of a ban would be tremendous, and that farmers were exposed to less of the chemical than the EPA was estimating—especially if they used new prepackaged containers of alachlor instead of dispensing it themselves.

"Oh yeah, that was the crux of the debate," says Joseph Reinert, a longtime EPA official who at the time was in charge of calculating the risks alachlor posed to farmers who spray it. "Since Monsanto made such a profit on alachlor—this was a very, very profitable chemical for them—the company came up with a lot of information."

"They spent millions of dollars doing studies," adds Diane Ierley, who was the manager of the EPA's alachlor review in 1984. "Under the law, the agency has to give the company the right to rebut the agency's concerns. We did, and they did."

Lapsley explains the process this way: "When you go out with your document and you propose to cancel specific uses because the benefits don't outweigh the risks, industry comes in and hits you on both sides. They tell you the toxicology isn't what you think it is and the exposure isn't what you think it is. And on the benefits side, they come in and give you higher estimates of what the product is worth to farmers."

The manufacturers' efforts sow enough uncertainty that the EPA tends to decide against bans even after initially proposing them, Lapsley adds. "There is a tendency against cancellation. We sure won't go in and cancel something if it might create a bigger problem. It's better the devil you know than the one you don't."

Monsanto executives say that they were indeed working hard to protect alachlor in 1984. By that time, they say, company scientists had carefully reviewed the cancer studies and all the other health-related research and were convinced that the weed-killer could be used safely. "This was an important product," says George Fuller, the director of product registration and regulatory affairs for Monsanto's agricultural division. "It was one that we had a very strong interest in defending. We thought it was the right thing to do, and we were going to leave no stone unturned in the defense process."

By November 1984, the EPA had given up on banning alachlor, agency officials say. Ranking EPA officials even rejected suggestions by Reinert and others to ban alachlor's use on a few minor crops where the chemical was not particularly important. "That way we would have looked like we cancelled something," Reinert says. "There were many people who argued that this was a likely human carcinogen, and we use millions and millions of pounds of it, so we ought to get rid of it." Instead, the agency moved to begin a special review of alachlor that would ultimately be concluded at the end of 1987 with a series of new but relatively minor restrictions.

But Canada would not be swayed. In February 1985, it banned alachlor, and the Canadian health minister later overruled a special review board that recommended reinstating the herbicide. The EPA's Lapsley believes that the two nations ended up making such drastically different decisions on alachlor because Canadian regulators operate in a different environment. Unlike their Canadian counterparts, EPA regulators under federal law must prove not only that alachlor is a health risk but that the chemical's risks outweigh its benefits.

"The Canadians tended to be more risk-averse than us," Lapsley says. "Their Health Canada [agency] would take some pretty aggressive positions, and they weren't bound by the same requirements in FIFRA [the federal pesticide law]. They just went out and did stuff. We had a heavy burden under the statute to show that the risk was unreasonable, and we knew we were going to be beat to death no matter what we did, and that we had to have a good case when we went out on a proposed ban. The Canadians didn't face that sort of pressure; they weren't bound by the same requirements."

The pattern is repeated over and over again in chemical regulatory battles: The government may start out with plans to move aggressively to protect public health, but the momentum quickly fades as regulators find themselves outspent and outlasted by corporations and hamstrung by weak laws and political pressure.

Having dodged the bullet of an alachlor ban in 1984, for example, Monsanto adopted a wear-them-down strategy, EPA officials say, as the company sought to defuse the lingering issue of possible contamination of lakes, rivers, and underground aquifers by alachlor. Testing by state officials in Iowa, Wisconsin, and elsewhere proved that the weed-killer was indeed showing up in all three sources of drinking water. As the EPA began its special review of alachlor, it was evident that the speed and degradability of the chemical as it moved through soil—its "environmental fate," in the parlance of the EPA—would be one of the key issues. The EPA already had in hand a study that showed that alachlor was in the water supply of Columbus, Ohio, and other midwestern cities, but Monsanto insisted that such examples were flukes. So the EPA designed a series of tests, to be conducted by Monsanto, that would show how quickly alachlor seeps through different kinds of soil and whether it breaks down into harmless compounds before it reaches drinking-water supplies.

What happened next reveals a fundamental flaw in the way the United States regulates pesticides, according to William Marcus, the senior science adviser in the EPA's Office of Ground Water and Drinking Water. Marcus, a well-known whistleblower, has been an outspoken critic of the chemical industry. For several years in a row, he says, Monsanto failed to conduct those leaching studies the way the EPA had asked—and the way the company had promised it would.

"What they [Monsanto] would do is they would reach an agreement with EPA about how they would submit the data and what kinds of experiments they were going to do, and then they would just submit material that was unacceptable," Marcus says. "And every time they did another year went by. They did this three or four years in a row. It was in Monsanto's best interests to screw it up every year, and they did, because every time they did it gave them another quarter-billion dollars in profit with alachlor. That is a basic flaw in FIFRA. As long as they're selling the product, it's in their best interest not to make their studies come in with adverse data."

"It did drag on," says Arthur Perler, a former EPA scientist who at the time was the chief of science and technology in the EPA's Office of Water (which has since been renamed). Alachlor's leaching potential "was the subject of many studies, and there would always be problems. I don't know that I would swear that the studies were designed to be defective. The information that's considered important to make an intelligent decision is changing all the time. But there were some institutional obfuscations."

Monsanto's Fuller maintains that the company never deliberately performed its leaching studies incorrectly. But he says that the EPA rejected at least one of the studies and that there often were misunderstandings about how these studies should be carried out because this was a new area for the EPA. "Yes, there were discussions back and forth, and we'd try to do the studies as best we could," Fuller says. "And I'm sure there were one or two that were rejected and we had to do over again. But it wasn't a question of a dispute. I think it was just a question of understanding what they wanted."

By 1990, Perler had left the agency. So had Stuart Cohen, who worked closely with him on the alachlor issue. The entire office had been reorganized, Perler says. The result, according to Marcus, was that the EPA stopped insisting that Monsanto comply with its protocol for the leaching studies. There was simply no one left in the office, he says, who knew enough to properly scrutinize Monsanto's research. "That's an institutional problem at EPA," Marcus says. "Because the companies draw it out, the chemical managers—the people responsible for following through—change, so the continuity is lost."

Says James Huff, a cancer researcher at the National Institute of Environmental Health Sciences, an arm of the National Institutes of Health: "A lot of the EPA people are administrators, not card-carrying scientists, and they're stretched thin. It's really hard to be at EPA and know all there is to know about every chemical you ever come up against. I'm glad I'm not working at EPA, because you get bombarded constantly by the industry side, and these industry guys make a career out of doing that and they're very persuasive. You get worn down, I guess."

By the early 1990s, the formaldehyde problem had become another one of these long-running sagas at the EPA, the kind some people dubbed "grazing buffalo" because they were big and never moved far.

Mark Greenwood, a partner in the Washington office of Ropes & Gray, a Boston-based law firm, took on the job of herding the grazing buffalo when he became the director of the EPA's Office of Pollution Prevention and Toxics in the Bush administration. Greenwood's staff had been demoralized by a 1991 federal court decision that threw out their years of work on asbestos, a known carcinogen.[6] If they could not prove that the health benefits of asbestos regulation outweighed the costs to industry, how could they ever advance regulations on other chemicals they believed were harmful, where the evidence (at least in industry's view) was not so clear?

Greenwood says one reason that he felt it was important to try to do something about formaldehyde was to prove that the Office of Pollution Prevention and Toxics was capable of action—or at least of bringing something to a conclusion.

The EPA first considered regulating formaldehyde in 1980, after the

Chemical Industry Institute of Toxicology's early rat study results. But formaldehyde manufacturers insisted that the findings were not relevant to humans. In one of the best-documented episodes of chemical industry politicking, the Formaldehyde Institute exploited the anti-regulatory philosophy of the Reagan administration's appointees at the EPA.

A thank-you note to a top EPA official from James Ramey, an executive of Celanese Corporation and the chairman of the Formaldehyde Institute, fell into the hands of a Democratic Congress eager to attack the Reagan administration for reported sweetheart talks with industry. The formaldehyde manufacturers and administration officials called these secret, off-the-record, improper, and possibly illegal meetings "the Science Court."[7] The two meetings were attended only by chemical company executives, officials of the Formaldehyde Institute, and officials of the EPA. There was never any public notice of the meetings or any official record of them, as the law requires.

"I found the forum intellectually stimulating and very helpful in putting a large volume of highly complex data into proper perspective," Ramey said in his letter. "Also, I applaud you for coming up with an innovative idea for encouraging open and objective dialogue between industry and government scientists. Many of us have long advocated a cooperative effort between industry and government in the regulatory process.... You have come up with a forum that works. I urge you to expand your idea to other chemicals and key necessary regulations based on good science. I predict that the 'Science Court' may be a lasting trademark of this Administration."

Ramey's forecast was on the mark. After Congress held hearings excoriating the administration for "the Science Court," Assistant EPA Administrator John Hernandez (the recipient of Ramey's thank-you note) was fired, Ramey was replaced as the Formaldehyde Institute's chairman, and the institute's chief lobbyist lost his job.[8]

Congressional hearings on the Reagan administration's meetings with representatives of the formaldehyde industry received some press coverage, but there was next to no follow-up coverage of the paralysis that plagued the EPA afterward.[9] Years passed before the agency would conclude, in 1987, that formaldehyde probably was carcinogenic to humans.[10] More lobbying followed: This time, the industry touted its new studies based on the mechanisms by which formaldehyde works.[11]

The EPA's Science Advisory Board questioned this work, but the agency essentially gave up trying to regulate formaldehyde as a cancer-causing agent.[12]

Still, there was a problem out in the real world. EPA officials heard about the lawsuits brought by people sickened by mobile homes in Texas; they knew, for example, of the $17 million verdict for the Pinkerton family in 1992 because of particleboard in their Kansas home. Nearly all scientists agree that formaldehyde gas can make people ill; could the agency entrusted with protecting the public do something about that, at least?

No, not very easily. Despite the many documented incidents of people becoming ill from formaldehyde gas, the manufacturers tenaciously argue that it is just not possible that formaldehyde makes people sick. It is not even possible, they say, that formaldehyde makes *some* people sick. "Various studies over a period of many years demonstrate that formaldehyde neither causes asthma nor has an effect on asthmatics different than its effect on persons who are not asthmatic," Borden, Inc., said in a written response to questions about formaldehyde's dangers.[13]

And through their main trade associations, the National Particleboard Association and the Hardwood Plywood and Veneer Association, Georgia-Pacific and other manufacturers of formaldehyde-glued wood products told the EPA that further testing was unnecessary. They argued that they had reduced the amount of formaldehyde in their resin mixtures since the late 1970s, when complaints first arose.[14]

"It's a completely different product today than it was 15 years ago," says Richard Margosian, the director of the National Particleboard Association. Margosian himself worked on formaldehyde reduction as an engineer at Boise Cascade Corporation in the 1980s before coming to Washington to work for the trade group.

Throughout the late 1970s and early 1980s, the formaldehyde and wood-products industries never admitted publicly that the gases from their products were a problem, but behind the scenes they struggled over what to do. One proposal frequently raised called for the fumigation of mobile homes with ammonia gas, which had been long known to act as a scavenger that could capture and neutralize formaldehyde in the air. But some of the technical experts believed that ammonia gas could damage the electrical systems of a home.[15] Other bad side effects of ammonia were not discussed at the meetings but were certainly known to the scientists there, ranging from ammonia's tendency to darken and damage wood-work to the fact that it is itself a noxious and potentially hazardous gas.

In 1979, one of the leading experts on urea-formaldehyde resins, Beat Meyer of the University of Washington, had published his findings that ammonia worked much better to reduce formaldehyde emissions *before* wood was cured—in other words, in the mill before it was shipped out.[16] The formaldehyde industry deemed this treatment too expensive. Eugene Skiest, an assistant to the vice president for science and technology at Borden, said that ammonia fumigation of mobile homes "might well be less expensive than improving products," according to the minutes of an October 1980 meeting of the Formaldehyde Institute. "A number of recent suggestions that fumigation be done at wood-products plants themselves were discussed. The economics appeared to be prohibitive, but fumigation of mobile homes themselves appeared attractive."[17]

In 1990, the Consumer Product Safety Commission issued a pamphlet on formaldehyde in which it warned consumers against this "attractive" solution. "We strongly discourage such treatment since ammonia this strength is extremely dangerous to handle," the pamphlet said. "Ammonia may damage the brass fittings of a natural gas system, adding a fire and explosion danger."[18]

According to Margosian, the formaldehyde and wood-products industries ultimately found a more permanent solution to the problem: They developed resins with less formaldehyde.

The National Particleboard Association points out that the CPSC has logged no complaints about formaldehyde fumes since 1991, and only a handful in the preceding years.[19] That compares to 1980, when the agency received 280 complaints from consumers, which the industry attributes to the fume build-up that foam insulation made from the old high-formaldehyde resins created in newly energy-efficient homes. (The commission logged one complaint in 1993 from a Michigan woman who said that she had difficulty selling a home with formaldehyde insulation.)

In a set of prepared responses to question about formaldehyde, Borden, Inc., also pointed to the lack of complaints to the CPSC, as well as the commission's ultimate decision not to regulate composite wood products or permanent-press clothing, as evidence that the substance is benign. "Formaldehyde emisssions from such products or permanent-press fabrics do not pose a risk of harm to human health and safety, either acutely or long-term," Borden said.[20]

But the CPSC also develops statistics on product hazards drawn from a nationwide survey of hospital emergency rooms. It estimates that at least 125 people had so much trouble breathing or such severe skin or eye irritation from particleboard or other wood products that they went to an emergency room in 1994 (the most recent year for which figures are available).[21]

Even if they have not called on the federal government for help, many people still believe that they have been hurt by formaldehyde in their homes. One can obtain some sense of the problem from a database maintained by the Association of Trial Lawyers of America. From 1990 to 1994, 30 lawyers reported having active cases on behalf of clients with illnesses they blamed on paneling or particleboard, mostly from mobile homes.[22]

Thad Godish, the director of Ball State University's Indoor Air Research Laboratory, who has written extensively on formaldehyde, agrees that the fumes from particleboard are "nowhere near as potent as they used to be." But he points out that formaldehyde is one of the strongest of chemical irritants: "It doesn't take a whole lot to cause mucous membrane irritation in sensitive people."

The EPA has determined that most people begin to feel eye, nose, and throat irritation when the fumes reach one-tenth of a part formaldehyde per million parts of air to three parts per million.[23] The industry insists that there is no evidence of irritation at one-tenth of a part per million "in the general population."[24] The emphasis is on those four words, because people who are sensitive, or allergic, to formaldehyde—estimated to be 20 percent of the population—feel symptoms at concentrations half that level (one-twentieth of a part per million), Godish says.

Since 1986, manufacturers have agreed to produce particleboard and other wood products that emit no more than three-tenths of a part per million. The program is monitored by their own trade group, the National Particleboard Association.[25] That is three times more powerful a formaldehyde level than the EPA believes causes irritation in many people. The EPA wanted to find out what happened in the air of a home that is paneled, floored, and furnished with various products that met the Particleboard Association standard; did the fumes dissipate, or did they concentrate?

"Perhaps it was naive," says George Semanyiuk, the EPA's lead staffer on the project. "We said, 'Let's go into homes and see what the levels are.' It was kind of simplistic; we didn't even know how we'd do it. But it had a purpose." The purpose was to take a serious look at the industry's claims of safety, considering the problems the EPA knew had been there.

At the meeting, the EPA's "simplistic" idea became a many-headed monster. Formaldehyde could be released from food in cooking or from the cabinets because of kitchen heat. It is used as a fabric finish in some draperies and permanent-press clothing. It is in some paints and household cleaners. It is even in cigarette smoke.[26]

In fact, however, the EPA had solid evidence that these other products contribute only minimally to indoor formaldehyde pollution. "I think there have been plenty of studies to show that those small sources contribute virtually nothing," Godish says.[27]

It took nearly a year and a half—and a threat from the EPA to obtain a court order against the industry—for the two sides to hammer out the details of a testing program in a document as thick as a small city phone book.[28] It was not what the EPA had wanted: a study of at least eight different homes—with people living in them—with different arrangements of paneling, flooring, and furniture. The industry agreed to spend $460,000 to test the air in one conventional and one mobile home, both uninhabited, with the furnishings changed several times to produce the different formaldehyde scenarios the EPA wanted.[29]

The EPA's budget includes virtually no funds to test chemicals; instead, Congress gave the agency power to order manufacturers to do the testing if the regulators can show evidence that a product poses a significant health risk. However, rather than get into a legal battle with the industry over whether the formaldehyde in wood products really was a significant risk—the agency had just lost its fight with the asbestos industry in court—EPA officials pushed for a voluntary testing program.

A funny thing happened, however, halfway through the tests. The particleboard and plywood groups, even though they are supported by huge corporations, ran out of money in the summer of 1995.[30] The attorney for the associations, Brock Landry, says that they could not pledge unlimited funds to a program that they never thought was necessary or feasible in the first place. Cost overruns had been $200,000. Landry adds that the associations ran into numerous difficulties, as they had predicted, in even finding test homes that real estate agents would agree to keep off the market.

The solution was to cut back the program even further, by abandoning testing in the mobile home. "Our argument has always been that a molecule of formaldehyde doesn't know if it's in a manufactured home or a conventional home," Landry says. If the point is to study formaldehyde, he says, it shouldn't matter where the scientists take their measurements.

It was a development that made the EPA officials sick at heart. Greenwood was no longer there; he had left the agency before the testing even began to join a Washington law firm. The EPA still had little leverage. It agreed to a testing program that would not look at the very type of home that its scientists knew always had the worst formaldehyde gas problems, whether the molecules knew it or not.

"Unfortunately, we got into a cost negotiation situation," Semanyiuk says. "I think the EPA would say today that we would prefer going with the [court] order approach, rather than voluntary testing, because we'd feel more comfortable with a little bit more of a legal framework there."

By March 1996, the industry had completed its tests, which showed that formaldehyde in the conventional home were one-fourteenth of a part per million or lower (0.076 ppm)—a level so low, the industry said, that humans could not even detect it. The EPA officials knew of evidence, however, that such low levels could irritate some people, and they wanted to test the results in any case.[31] The EPA officials suggested an idea to assuage their doubts; they said they wanted the report sent to several outside scientists for "peer review." Landry says the industry viewed this as just another unforeseen responsibility and expense foisted upon associations that had, despite their doubts about the whole concept, been "fully cooperative."

The particleboard association's board of directors voted not to finance the peer review. The EPA began to make plans to go ahead and recruit outside scientists to review the validity of the study anyway. And Semanyiuk says that the matter is not over. The EPA viewed the one-house study only as a pilot, to assess the methods the agency had developed for testing. After peer review, the agency planned to call the industry associations together again for another meeting, to negotiate what to do about the possible hazards of formaldehyde, still one of the most prevalent chemicals in the American home.

Godish criticizes the EPA's penchant for "taking a very simple task and making it much more complicated than necessary," adding: "EPA projects are horribly expensive. They have put in what are called standards. You have to have a quality assurance program and a lot of additional testing. Well, when you have lots of money, you can do a nearly perfect study. With less money you have to do less and say, 'These are the limitations.' Some data is a lot better than no data. When I've looked at some of their protocols for testing on indoor air quality, I think, 'My God, you put all this in and it'll sink the Republic.' What EPA wants is their cake and to eat it, too. They want this information, and they want industry to pay for it."

Yet this guarantees that public health officials and other federal regulators will have no control over studies of toxic chemicals. "Congressmen and the industry cry about science and risk assessment," Godish says. "There is no chemical I know of that the science is so strong to implicate and support regulation as formaldehyde, and there has been no regulation from the EPA."

On the other side, the National Particleboard Association's Margosian says that the products made by the group's members have been unfairly stigmatized for years. He adds that magazine, newspaper, and television reports routinely mention formaldehyde among the top indoor air pollutants, even though the industry does not agree that it is—at least today—a significant problem at all.

"It strikes me we've gone through 16 years and three federal agencies," says Brock Landry, the association's attorney. "It might be an appropriate time to step back and say, 'We've got an awful lot of data, now. We have enough information to conclude this.'"

With the dry-cleaning chemical perchloroethylene, federal officials likewise have become mired in the hopeless mission of placating different interest groups. When new and frightening evidence of perc's dangers appeared in the early 1990s, it should have spurred the EPA to quick action. Instead, it only prolonged the agency's paralysis on this long-recognized hazard.

Judith Schreiber, a toxicologist with the New York Department of Health, had set out to study fumes from dry cleaners after realizing that her agency was receiving numerous complaints from upstairs apartment dwellers about the sickening sweet smell of perc. On one visit, in "a typical infant's bedroom, with a crib and lacy things," she recalls, she found levels of perc fumes higher than the rather liberal limit permitted in the workplace. On another visit, an apartment 12 stories above a dry-cleaning shop had perc 60 times the workplace limit. The chemical had made its way upstairs through heating ducts and vents.[32]

The huge perc concentrations translated into staggering cancer risks. The risk for people who worked in dry-cleaning establishments was the highest: a 1 in 10 risk of developing cancer. For children in apartments above dry-cleaning establishments, the risk was also enormously high, from 1 in 1,000 to 1 in 10,000. Even the risk for those exposed to perc in the outdoor air was substantial: as high as 1 in 100,000.[33]

Considering that the EPA's policy is to step in and take action with cancer risks even as low as one in a million, the New York dry-cleaning numbers screamed for action. Yet, in the upside-down world that chemical regulation has become, the magnitude of the crisis virtually ensured that nothing would happen. One worry was that the Natural Resources Defense Council, a not-for-profit organization with a long history of going to court to force the EPA into tougher enforcement of the law, and other environmental groups would criticize federal officials for not having learned about the danger and acted on it earlier.

"I am very concerned that when word gets out on these numbers, even though they are preliminary and need review, NRDC et al. will go ballistic," Mary Ellen Weber, the director of the economics, exposure, and technology division of the EPA's Office of Pollution Prevention and Toxics, wrote in a memo to Mark Greenwood, the director of the office. "The timing is unfortunate since we can be asked why we didn't know how bad perc... exposure risks were until now, after talking about perc for years."[34]

EPA officials were also worried about what they knew they would hear from the lobbyists: that action to curb perc, or even rumors of action, could have a devastating effect on the dry-cleaning industry. That would be an especially stinging rebuke when the EPA had been trying to quell such criticisms by working with industry in voluntary pollution-reduction programs.

Dry cleaning, in fact, was the first industry the agency had thought it could nudge gently into doing business with fewer chemicals. In the EPA's new "Design for the Environment" program, a group of agency economists and scientists had been working to show that "dry" cleaning with water was possible and economical, as long as cleaners used machines that carefully regulated the temperature, agitation, and drying of the clothing.[35] But the International Fabricare Institute and the Neighborhood Cleaners Association turned out to be as immovable as the solvent makers with which they worked closely—Dow, PPG Industries, Inc., and Vulcan Materials Company.

The uncertainties that are inherent in the EPA's science can be added to this mix of pressures. Human studies show that the risk of developing urinary bladder and esophageal cancer "may be increased by employment in dry cleaning," according to the International Agency for Research in Cancer, which evaluated the research available as of 1995. But that is a far cry from showing that one out of 10 dry cleaners is dying of cancer, as the EPA's math based on rodent studies would predict. But the International Agency on Research in Cancer viewed the uncertainties as great enough that it took the unusual step of ruling that dry cleaning entails only possible carcinogenic exposures, at the same time it ranked perc as a probable carcinogen.

The decision that the risk numbers for perc would never see the light of day came slowly and painfully. Greenwood gathered the EPA's toxics staff soon after hearing the figures, a meeting he would later tell Weber "went well," although the disagreements among the staff presented a "difficult issue." "While nobody seemed willing to speak up," he said, "I had the sense that some staff do believe we have a quasi-emergency situation."[36]

It was difficult to restrain the anxiety that was building over perc. Memos were exchanged. What were the greatest uncertainties in the science? Extrapolating from animals to humans? Gauging whether the risks rose steadily with the dose? Knowing the mechanisms by which perc worked?[37] Perhaps a different office within the agency should redo the assessment.

The debate continued for months, and the agency decided that it could not release its long-overdue final paper on the benefits of alternatives to perc dry cleaning, the Cleaner Technologies Substances Assessment, until it had finalized the risk numbers. Without revealing what the preliminary numbers showed, EPA officials, in telephone calls in early August, informed all the groups that had been working on the dry-cleaning project of the delay.

"The enviros and labor people were unhappy about delaying release of the hazard and risk pieces until the technology pieces could be integrated and industry was delighted," Weber said in a memo to

Greenwood.[38] One of the most unhappy people was Bonnie Rice of Greenpeace. She kept pressing the agency for the risk figures, and Greenwood told Weber to call her again so that she would not be able to portray the agency as failing to respond to her inquiries. Weber soon reported cautious optimism that Rice would hold off on making their lives hard: "She is not happy and I expect she may elevate this, but the conversation was polite, if not warm, and, on balance, I think she is slightly mollified!"[39]

After numerous internal meetings to discuss the options, none of which seemed attractive, EPA officials began to gravitate toward an unusual solution. They would release the numbers, but the numbers would not resemble any cancer-risk numbers the EPA ever had released before.

The numbers would look a bit like a pyramid, with more zeroes at the end of every succeeding row. They would give people a sense that workers were more at risk than people who lived in apartments upstairs, and that apartment dwellers were more at risk than the public at large. But nowhere would the EPA spell out exactly how big the risks were. Any risk scientist or mathematician could look at such a chart and know immediately how to calculate the "slope factor" that, with some more calculations, would lead to the same stunning individual risk numbers that the EPA had arrived at months earlier: 1 in 10 for dry-cleaning workers, 1 in 10,000 for the public at large.[40] The EPA never translated the numbers into plain English that the public could understand.[41]

Weber's misgivings about the September 22, 1994, decision to use a "margin of exposure" chart had only to do with the pressures that she knew the agency would face from labor and environmental organizations on the inevitable delay to create the new calculations. "The stakeholders have been clamoring for the report's release for some time and its delay contributes to their perception that we are hiding something," she said in a note to Joe Carra, one of her deputies. With dark humor, she enclosed in the envelope an eight-year-old report on the dangers of perc. "In case you are feeling a little sentimental, I've attached a January 1986 report on PCE [perchloroethylene]," she wrote. "It seems that some projects just never die."[42]

In December, with the work on the numbers nearing completion, EPA staffers again became gripped with nervousness. William T. Waugh, the deputy director of the chemical screening and risk assessment division in the EPA's Office of Pollution Prevention and Toxics, wrote a memo to other members of the team. He was worried that the group had been concentrating solely on completing the document, without thinking enough about the outcry that could follow.

Waugh then proceeded to spell out the problems that the EPA had discovered with perc—a stark description the public would never hear: "Nursing babies whose mothers work in dry-cleaning shops are exposed to PCE 20 to 30 times higher than the EPA considers safe.... Just living in a building with a dry cleaner downstairs results in exposures to babies above the level EPA considers safe.... The average person can be exposed to levels dangerously close to the RfD [the reference dose, or danger threshold] from clothes they bring home from the dry cleaner.... EPA's analysis shows that *anyone* who works in a dry-cleaning shop is at considerable risk. Workers are exposed to levels of PCE hundreds to thousands of times higher than is considered safe....

"My sense is that the PCE risk characterization will be released to the press as soon as it goes to stakeholders (if not sooner). I suggest [that we] clearly articulate our message as it relates to the risks of PCE and what we are doing about it. Write it down. Read it out loud. Listen to it. Make sure it passes the laugh test in the light of the black-and-white risk numbers in the characterization."[43]

In a later memo, Waugh would add that the seven-page fact sheet the EPA had drafted to explain the risk characterization "falls short of what we need to prepare for public dissemination," commenting, "My sense is we are running out of time."[44]

Waugh was wrong on that point, however, because the EPA was far from releasing the risk numbers. After the New Year, in 1995, pressure began in earnest from the dry-cleaning industries. The watered-down "margin of exposure" expression was not to their liking. They wanted the Cleaner Technologies report to have no figures on the risks of perc at all.

"Media attention would needlessly alarm the industry's employees and customers and result in significant economic harm to an industry composed primarily of family-owned-and-operated and small businesses," said a letter to the agency that was signed by representatives of Dow Chemical and a host of organizations representing solvent and equipment makers, franchisers, and the dry cleaners themselves.[45] The industry was requesting a meeting with Carol Browner, the EPA's administrator, over the issue, and members of Congress also had begun calling the agency to complain.

Meanwhile, the International Agency for Research on Cancer, meeting in Lyon, France, concluded that because of new studies of workers and residents, it no longer considered perc merely a possible carcinogen but a probable one.[46] Eric Frumin, the Amalgamated Clothing and Textile Workers Union's health and safety specialist, called Ohad Jehassi, an EPA economist who had taken the lead on the Design for the Environment project, to ask whether the agency would follow suit. "I don't know what to tell him," Jehassi, clearly disheartened, wrote in a memo to Weber.[47]

Lynn R. Goldman, the EPA's assistant administrator for prevention, pesticides, and toxic substances, met with representatives of the dry-cleaning industry and environmental organizations. She asked them for input on the method the agency should use to express cancer risks. By August 1995, a year and a half after the EPA first calculated the alarming risk numbers, the letters and comments were in. Nothing had changed; the EPA still was caught amid opposing—and seemingly irreconcilable—pressures.

"Almost uniformly, industry does not support any approach for the characterization of risk... at most suggesting the presentation of exposure information only," William H. Sanders, who by then had succeeded Greenwood as the director of the EPA's Office of Pollution Prevention and Toxics, said in a memo to Goldman summarizing the comments. "Conversely, the public-interest groups almost exclusively advocate the use of the [numbers that explain the risk to individuals]."[48]

The EPA would never formally quit its work on dry cleaning. But sometime in the fall of 1995, the project slipped quietly into hibernation. All of the work that the staff had done now lay beneath the warm haunches of the big bureaucratic bear—a sleepy state that pleased the industry. Labor and environmental organizations could rouse the issue for perhaps a few more meetings, another round of comments or two, but without any real action, there was nothing more they could do.

Bill Hayes, Dow's director of product stewardship, whose job is to promote the safe handling of chemicals once they leave the factory, says that the risk numbers the EPA developed were a "worst-case scenario" that did not deserve to see the light of day because they were based on faulty science. "They take positive animal studies, and combine them with equivocal studies of people—studies that were, in fact, negative or had small statistical increases," he says. "EPA guidelines assume no threshold [no low, safe level of exposure] and they also assume that man is more susceptible than animals, when in fact the opposite is true."

Steve Risotto, the director of the Center for Emissions Control, an arm of the halogenated solvents industry, makes a further point—that the EPA will lose credibility among the people it is seeking to protect by releasing the numbers: "What happens when you tell dry cleaners that one in 1,000 of them are dying of cancer? They're going to look around and say, 'Where are the people dying?' These numbers don't coincide with what they know. They're going to say, 'These guys are full of it.'"

Not everyone in industry is unsympathetic to the EPA. "To me, the regulators are in a no-win situation," says Manfred Wentz, the vice president for environmental affairs at R. R. Street & Company of Naperville, Illinois, which markets perc produced by Vulcan. "On the one hand, they have industry advocates, and on the other, those with an idealistic approach. And whenever the EPA makes a little bit of compromise or goes in the direction of accepting uncertainty, they are criticized by the other side."

Wentz, who began his career at the Fabricare Institute and spent 14 years in academia, including a stint as the chairman of the University of Wisconsin's department of textile sciences, says he "firmly" believes that the EPA's method of gauging perc's risks to man from animal are flawed. But he is also a strong advocate of working on alternatives to perc. His company even produces soaps for the professional wet-cleaning alternative to dry cleaning that it currently sells only in Canada. "Some people in industry have labeled me a 'green' man," Wentz says. He believes that a polarized debate will only create a backlash that will make it more difficult for the industry to give up perc.

"American entrepreneurs don't want to be told by anyone how to run their business, let alone by Greenpeace," Wentz says. "These sides need to be constructively brought together. I firmly believe in dealing with the issues factually, rather than politically." He says he felt that the EPA's Design for the Environment program was "a golden opportunity" for such a dialogue.

Greenwood says now that the EPA should not have spent so much energy battling over the expression of cancer risks. Instead, he believes that the EPA should have moved forward with its study on alternatives to dry cleaning—a study that now, ironically, may never be released.

Putting On the Pressure

Sometimes, delay alone cannot wear down the regulators. In those cases, the chemical industry is well equipped to rally a variety of troops into action: its lobbyists and economists, sympathetic small-business customers, and even other government agencies that the environmental officials cannot ignore. "Whenever you're talking about a heavily used, very important pesticide, there's always political pressure to keep it on the market," says the EPA's Reinert, the senior policy analyst who worked extensively on the alachlor issue. He recalls traveling to Monsanto's hometown of St. Louis to chair a meeting of the American Chemical Society and taking a tour of a Monsanto manufacturing plant. "I remember when we were on that tour they told us that if EPA cancels alachlor, two-thirds of the people who work here will lose their jobs," he says. "And there were editorials in the newspaper every day defending alachlor during this meeting."

A rare window into a chemical industry war room has been opened because of minutes of Formaldehyde Institute meetings that have been made public in lawsuits. Lawyer S. John Byington, a former chairman of the Consumer Product Safety Commission—the very agency that was giving manufacturers trouble at the time—gave the manufacturers a checklist on how to handle his old agency, which was preparing to ban formaldehyde foam insulation after receiving some 9,000 complaints from consumers.[49]

"We are faced with both a health and political problem," according to the minutes of the meeting. "[Byington] urged the institute to break down the defense into the technical, legal, and political areas. Mr. Byington listed the following suggestions, stating that taking the offensive is now critical.

"1. The Institute must get the facts and force the agencies to deal with facts. We must refute their data if they cannot supply backup materials.

"2. He urged the Institute to use the administrative procedures act of all agencies. . . . He especially noted Section 10 of the [Consumer Product Safety Commission] Act. [These laws allow outsiders to sue the government, among other things, for arbitrary and capricious actions.]

"3. Use the agency's personnel. It is necessary for Institute to work very closely with all agencies and find people within the agencies willing to work with industry.

"4. Use the legislative mechanism. Attend the appropriations and oversight hearings, raise questions at all of these hearings.

"5. Be prepared to go to court. If the agency does not believe you are willing to take this final step, half of the input is usually put aside. Legal counsel should be present and identified at any meeting with an agency."[50]

Nancy Steorts, now a business consultant in Dallas, was President Reagan's appointed chairwoman of the Consumer Product Safety Commission at the time that the Formaldehyde Institute's assault began. She still seethes at the memory because, although she proudly describes herself as a pro-business Republican, she believed that there was a real problem with formaldehyde foam insulation that the industry was not willing to face.

"We gave the industry, frankly, every benefit of the doubt," she says. "We suggested labeling and information programs—doing something for the consumers who were having horrendous problems and had to repair their homes. The industry was not willing to do this. They got every law firm in town to work to convince the commission it wasn't a problem. They used every tactic possible, they even formed a coalition to try to destroy the CPSC. It was absolutely the worst gutter tactics. I never saw anything like it."*

The Formaldehyde Institute sued and, by carefully selecting in which court to bring the case and presenting three scientific studies paid for by manufacturers, obtained a court order striking down the Consumer Product Safety Commission's ban.[51] However, by then, enough consumers had learned of the dangers of foam insulation that the market

---

*Steorts believes that indoor air pollution is a serious problem, based partly on her own experiences. In the spring of 1995, she was forced out of her home in Dallas for nearly three months because of her reaction to newly installed carpeting. She became unable to breathe and went into convulsions. She believes that the culprit was carpet padding—scrap fiber, bound together with glues, much in the way that scrap wood is bound together to make particleboard.

for the product had dried up. Steorts is gratified that her agency informed consumers, and she believes that one of her most important achievements was persuading real estate agents to disclose the presence of foam insulation whenever they make a sale or rental deal.

Formaldehyde remained a serious problem in homes, however, because the gas from particleboard and other pressed-wood products was far stronger than the chemical fumes from the foam insulation. Godish, the director of Ball State University's Indoor Air Research Laboratory, argues that the government agencies were quicker to act on the foam products because their effects were more dramatic and the source more obvious.

The Consumer Product Safety Commission, after suffering the brunt of an industry attack in the foam insulation battle, took a long time to address particleboard. The agency's chief scientist on formaldehyde, Peter Preuss, had left for the EPA. Steorts's term had ended. Ultimately, after numerous meetings with industry representatives, the commission decided that the steps taken by the industry were sufficient.[52] The wood-products industry today protests correctly that the EPA is forcing it to deal again with a problem that the CPSC already views as settled.

The story of the EPA's escalating estimates of alachlor's economic benefits is yet another example of how deferential the agency can be to manufacturers when confronted with lobbying pressure. When the agency first proposed restrictions on alachlor in January 1985, it estimated that lower crop yields and higher production costs would add $157 million to the cost of agricultural production during the first year of an alachlor ban. The overall cost to society would be $330 million to $340 million during the first year, the agency estimated, because prices would rise owing to declining yields per acre.[53]

But then Monsanto went to work. It flew in farmers and consultants who extolled alachlor's benefits at meetings with the EPA, and it arranged for dozens of agricultural scientists from around the country to write letters arguing that the chemical was far superior to its main alternative, metolachlor, a herbicide made by Ciba-Geigy. If alachlor was banned, Monsanto argued, food production costs would rise $330 million to $430 million, not $157 million as the EPA had argued.[54]

By the time the EPA completed its review of alachlor in October 1986, proposing a series of relatively minor restrictions instead of the ban it had initially planned, its assessment of alachlor's benefits had changed dramatically. The EPA, in fact, had gone even farther—a lot farther—than Monsanto had asked. A ban, the agency now concluded, would add $510 million to $759 million in food production costs the first year, at a net cost to society of $302 million to $508 million. Even after ten years, the EPA said, the ban would still raise production costs by more than $125 million annually—and much higher than that if enough farmers did not accept metolachlor as an alternative.[55] The EPA later downgraded its estimates slightly, putting the first-year production cost at $413 million to $465 million—and the cost to society at $300 million. But the agency still ended up with an estimate that was well above Monsanto's.[56]

In accepting Monsanto's explanation of alachlor's economic virtues, the EPA not only discarded its initial estimates, it also rejected a detailed analysis by two economists at the Agriculture Department. Their mid-1986 report concluded that farmers would readily adapt to metolachlor and other less toxic alternatives to alachlor. Production costs would actually *decrease* $64 million in the initial year of an alachlor ban, and the overall cost to society would be just $124 million. After five years, the production surplus would be $186 million, and the overall cost to society would be $149 million.[57] Even before the USDA study was finished, Monsanto executives were seeking a meeting to try to persuade the report's authors to change their conclusions. "Monsanto was not happy with what we came up with," recalls USDA economist Craig Osteen, the lead author of the 1986 report. "They came over and talked to us. I think they would have liked for us to go through and change the report. We wouldn't do that."

Even the USDA study, though far tougher on alachlor than the EPA's assessment, may have drastically underestimated the benefits of removing alachlor from the market. It ignored the fact that more corn was being produced than the market could absorb, which meant that taxpayer-funded price supports were subsidizing farmers because prices were low. In their meetings with the EPA, Monsanto consultants hammered home the message that an alachlor-treated cornfield yields at least four more bushels per acre than a similar field treated with metolachlor. But even though surplus corn was rotting in silos, the EPA never questioned whether the extra yield was really desirable.

Ever since the Depression, the federal farm subsidy system has been based on the premise of holding down supply to keep prices from plunging. (Most such subsidies will be phased out over the next seven years under legislation that Congress passed in 1996.) Farmers are often paid to let their land lie fallow and to send surplus crops into storage, where they must be fumigated—all at taxpayer expense. If prices fall anyway, despite the efforts to keep supplies down, taxpayers pay farmers the difference between the market price and a government-established target price. But the EPA, in evaluating pesticides, takes the opposite view—the industry view. It simply assumes that bigger yields are better, no matter what happens to prices. "We are never willing to face up to that," says Joseph Reinert, the EPA policy analyst who worked extensively on the alachlor issue. "How do you put a value on field corn in this country when we pay people to limit how much they store and we pay people to fumigate it while it's stored? In this country, if the yield goes down a few bushels per acre, who cares?"

In the midst of the alachlor controversy, Maureen Hinkle, the National Audubon Society's agricultural program director, made the same argument to the EPA, but to no avail. "What real benefits were there?" she recalls asking. "We had surplus corn. I mean, we were paying through the nose to build extra storage."

Today, some water utility officials say they fear that the EPA will show the same tunnel vision in its estimates of the financial impact of banning or severely restricting atrazine, which is currently under agency review. They worry that the EPA will not sufficiently account for the potentially huge cost of removing the chemical from drinking-water supplies. Taxpayers do not have the luxury of ignoring those costs. The small town of Higginsville, Missouri, for example, has spent more than $50,000 trying to remove atrazine, and in mid-1996 it was preparing to raise rates for the local water district's 9,000 water customers—probably by about 7 percent, or $21 a year, for a typical customer—to offset the cost of removing enough to meet EPA safety standards, which were to take full effect in 1996.

Across the Midwest, in big cities like St. Louis and Omaha and small towns like Higginsville, water suppliers have been grappling with the same problem. A study by the American Water Works Association estimated that at least 193—and perhaps as many as 823—water utilities will not be able to meet the EPA's atrazine standard and will have to spend anywhere from $51 million to $216 million a year to come into compliance.[58] The bill could be even higher, because Des Moines, Kansas City, St. Louis, and several other cities in the Midwest have reacted to public concern over atrazine by pledging to completely eliminate it from their water systems, even though levels in those cities are already low enough to comply with EPA rules.

The EPA fails to consider other indirect costs, such as the medical costs to farmers and consumers who are injured by farm chemicals. Other uncounted costs are associated with the changing farm practices that have come with heavy herbicide use. With more than 95 percent of the nation's corn crop now sprayed with herbicides, most farmers have abandoned cultivation and crop rotation—practices that helped keep insects in check. As weed-killer use has increased, crop losses to insects have jumped from 3.5 percent to 12 percent of the average cornfield since the 1940s even though insecticide use on corn has increased more than a thousandfold.[59] That is why David Pimentel, a professor of insect ecology and agricultural science at Cornell University, thinks that herbicides are responsible for a major portion of the nearly $200 million a year that farmers spend on insecticides.

Finally, there is the question of whether the bigger-is-better philosophy of agriculture that heavy pesticide use demands is really in the best interest of rural communities. Chemical-based farming has driven yields up and commodity prices down, and that has inevitably led to big farms getting bigger while small farms disappear, argues John Ikerd, an Agriculture Department economist affiliated with the University of Missouri. He adds that analyses of a pesticide's economic impact, such as those studies conducted by EPA, fail miserably to account for the full cost to society.

"There's no meter that racks up costs when you put atrazine in a stream and somebody has to put in a treatment system downstream to clean it up, or when a rural town is no longer viable and small farmers have to go on welfare or move somewhere else and depress the wage market elsewhere," Ikerd says. "There's nothing in the market that accounts for that."

The EPA's industry-friendly way of calculating the benefits of pesticides leaves outsiders feeling frustrated and powerless. "We're paying subsidies to these guys [farmers] to grow corn, yet we don't want to force them to pay more for [pesticide] products," says J. Alan Roberson of the American Water Works Association. "Our guys [who operate water utilities] have to deal with the consequences, and we get nothing from the government. I don't know how effective we are at changing the EPA's opinion. We have eight people in this office, but the companies have got lawyers and consultants and everything else. We're way outgunned."

Sending in the Surrogates

Some of the most effective attacks on regulatory agencies come not from the companies themselves but from industry's allies in academia and government. Often these industry allies occupy positions within the bureaucracy or on outside advisory boards where they can apply heavy pressure on regulators.

Consider the case of James Swenberg. He was accustomed to undertaking research projects on behalf of manufacturers as the director of the department of biochemical toxicology and pathobiology at the industry-financed Chemical Industry Institute of Toxicology, where he worked throughout the 1980s. In June 1986, for instance, he reviewed a rat study at Monsanto's request—he was not paid to do it—and concluded, as the company had hoped he would, that the brain tumors that alachlor caused in laboratory rats originated as nasal tumors and had metastasized to the brain.[60] His findings helped to persuade the EPA not to add brain tumors to the list of alachlor-induced cancers in rats, which already included nasal, stomach, and thyroid tumors.

Just five months later, Swenberg was in a position to give Monsanto an even more valuable assist. At the same time he was working for the institute, he was also a member of the EPA's eight-person Federal Insecticide, Fungicide, and Rodenticide Act Scientific Advisory Panel. The panel had sweeping authority to question—but not overturn—EPA decisions, including whether to restrict or cancel a pesticide. During the late 1980s, the panel did not hesitate to use that authority and repeatedly accused the EPA of overestimating health risks—most notably with the Alar case, in which it persuaded the EPA that more studies were needed, causing a five-year delay in the EPA's proceedings to cancel the apple pesticide. Later, Senator Joseph Lieberman, a Democrat from Connecticut, would reveal at a congressional hearing that "seven out of the eight members of the SAP...that stopped the move to ban Alar in 1985 worked for the chemical industry at the same time they were on the panel."

As it did in the case of Alar, the Scientific Advisory Panel quickly adopted the chemical industry's position in the alachlor case. By 1986, Monsanto had conceded that alachlor caused nasal, stomach, and thyroid tumors in rats, but there was controversy over studies that suggested the herbicide also caused lung tumors in mice. The EPA asserted that the number of lung tumors was well above the expected number, but Monsanto argued that the number of tumors was not unusually high for the particular strain of mouse being tested, the CD-1. It was a crucial issue, because the appearance of tumors in a second species made it much easier for the EPA to classify alachlor as a probable—not merely possible—human carcinogen and strengthened the case for a ban on alachlor.

When the advisory panel took up the question of the mouse tumors at a morning-long meeting on November 19, 1986, James Swenberg dominated the questioning. He interrogated EPA scientists so aggressively that at one point one of the scientists, Judith Hauswirth, told Swenberg, "You are very interesting to deal with," prompting laughter from the audience.[61] At the end of the meeting, the other members of the advisory panel decided to adopt Swenberg's position, concluding that Monsanto was right about the lung tumors and that the EPA was wrong. While the EPA ultimately refused to change its classification of alachlor as a probable (B2) carcinogen, the uncertainty over its carcinogenicity in humans has been a key factor in the agency's decision to allow alachlor to stay on the market.[62] Today, the EPA still cites the advisory panel's 1986 conclusions as evidence that the cancer risks of alachlor may be overstated and that it may not cause cancer in humans at all. The panel's conclusions were also a key element of the agency's decision nine years later to adopt a new way of calculating alachlor's cancer risk that has the effect of playing down that risk.[63]

Reto Engler spent years working with Swenberg and the rest of the advisory panel when Engler was a senior science adviser in the EPA's pesticides office. "I don't think he is tainted by doing work for industry, but he is a very one-side-of-the-coin type of scientist," said Engler, who is now a private consultant. "He only believes in a carcinogen beyond any reasonable doubt. We all know this is a gray area, but his gray is practically nonexistent. And the line for Doctor Swenberg between black and white is somewhat different than the line is for other scientists."

Swenberg says that he remembers asking EPA scientists tough questions about the mouse tumors. "I think my questions were thorough rather than aggressive," he says. "I think I'm very objective, but I do hold them to the science. The EPA has a tendency to do a sloppy job on statistics, and that's what happened with alachlor." Swenberg left the EPA advisory panel when his four-year term expired in 1990. He will never serve on the EPA board again, he says, because agency officials were too slow to speak up in defense of panel members' integrity when Lieberman and others raised conflict-of-interest charges after the Alar controversy. "I was very upset," he says.

Swenberg is now a professor at the University of North Carolina and the director of the university's toxicology program. Since 1995 he has also been a paid consultant to Monsanto, but Swenberg and company officials say that he was never paid by Monsanto during the four years he served on the EPA advisory board. Swenberg says he sees nothing wrong with industry-financed researchers serving on agency panels, although he acknowledges, "I guess if you're an uninformed outsider that could look strange."

Because the EPA's advisory boards do not vote but instead arrive at consensus documents—reports that each and every member agrees on—industry officials can exercise a powerful influence even if labor, consumer, and environmental organizations are also represented. Often, these "consensus" documents end up containing statements to mollify every diverse interest at the table.

For example, as recently as 1996, the perchloroethylene industry was citing the 1988 document produced by the EPA's Science Advisory Board—another committee of outside scientists—as stating that "there is insufficient evidence that perc is a human carcinogen." The perc industry's reading of this document, however, is decidedly selective. Actually, the board had said in its March 9, 1988, transmittal letter to then–EPA Administrator Lee Thomas that perc was on a "continuum" between possible and probable carcinogen, noting that the distinction between the two classifications "can be an arbitrary distinction."

"From a scientific point of view, it seems inappropriate for EPA and other agencies to regulate substances that are classified as [probable carcinogens] and not to consider regulation of compounds classified as [possible], regardless of the level of human exposure," the board said. Some very highly carcinogenic substances may not need regulation, because not enough people are exposed at a high level, the board added. Conversely, a substance classified as a possible carcinogen may be a much greater threat because exposure is high.[64] The industry never mentioned this statement, but the high level of perc use and, in some places, exposure is exactly why many in the EPA believed that regulation was necessary.

Although the chemical industry often criticizes "big government" when dealing with environmental and health officials, the industry frequently uses to its advantage the fact that different government agencies have responsibilities for environmental protection and public health.

Bureaucrats at the U.S. Department of Agriculture, for example, were not quite sure how to react in the summer of 1985 when Monsanto executives made an unusual request. The company had just completed a massive 20-volume report for the EPA that covered alachlor's toxicology, its leaching ability, and just about every other alachlor-related subject that the EPA was investigating as part of its special review. But now Monsanto wanted the Agriculture Department to review the report, too, and to plead the company's case before the EPA. "It is critical that EPA recognize the importance of this decision to American agriculture," Nicholas L. Reding, an executive vice president of Monsanto, wrote in a June 1985 letter to Agriculture Secretary John Block. In the letter, Reding thanked Block for meeting with Monsanto executives a few days earlier and for "your promise to review the matter of USDA involvement."[65]

Federal pesticide law specifies a clear division of labor between the two agencies. The EPA is in charge of reviewing health and environmental studies and has the final decision on how pesticides are regulated. The USDA's role, traditionally, is that of an interested bystander; its opinions are not binding on the EPA and are almost always confined to how farmers will be affected by a proposed EPA regulation. As part of its strenuous efforts to defend alachlor, however, Monsanto executives were suggesting that the USDA assemble a panel of agricultural scientists to review the company's entire 20-volume report and then to "intercede on their behalf to EPA," as Charles Smith, the Agriculture Department's chief pesticide liaison to the EPA, wrote in a June 17, 1985, memo. "The downside of this would be twofold. If we disagree with Monsanto's analysis, it would not be helpful to them. . . . On the other hand, if we agree with Monsanto and communicated that to EPA, we would be going to a sister federal agency on behalf of industry. Something we have not done in the past. Further, we would most likely alienate one company (Ciba-Geigy) in our support for the other."[66] (Ciba-Geigy manufactures metolachlor, which at the time was alachlor's chief competitor.)

In an interview, Smith recalls that one USDA official, whom he would not identify, was particularly enthusiastic about getting involved on Monsanto's behalf. "There was some guy at Agriculture who was looking to get a job with industry," Smith says. "It was his position that we ought to look at Monsanto's report. But that would have opened up an accusation from EPA and any other number of organizations that the Department of Agriculture was in bed with industry. It was part of my job to make sure that didn't happen."

Monsanto's Fuller says that the company asked the USDA officials to review the 20-volume report so that they would be comfortable defending alachlor to the EPA: "By giving them information on safety, even though that's not normally what they would review, we wanted to give them the information that we felt would provide the assurance that we were not asking them to go out on a limb."

Ultimately, the USDA decided not to formally review Monsanto's report. But scientists within the department and throughout the university land grant system flooded the EPA with testimonials extolling alachlor's virtues. Smith, now a fundamentalist Baptist pastor in Washington, makes no secret of what he and his USDA colleagues thought about the EPA's efforts. "We felt that EPA was always overdoing it in the toxicology and the safety," he says. Man-made chemicals, Smith argues, are no different from naturally occurring toxic chemicals that have been around for millions of years. Asked what he thought about the argument that the human body has had millions of years to evolve ways of metabolizing natural carcinogens but only a few decades to develop defenses against synthetic chemicals, Smith says, "I do not believe in evolution, both from a biblical and a scientific standpoint."

Not all USDA officials reject evolutionary theory, of course, but the department views pesticides in a fundamentally different way than the EPA does—a difference that chemical manufacturers have repeatedly sought to exploit as they try to stave off EPA regulation.

The Agriculture Department's pro-industry tilt was also an issue seven years later when the USDA's working group on water quality issued a report on atrazine. Its report "seems to minimize the potential problem, not only with utility compliance, but with the protection of public health," John Sullivan, the deputy executive director of the American Water Works Association, would complain in a letter to congressional staff members.[67] A few days after Sullivan wrote the letter, the USDA convened a two-day conference on atrazine contamination of water supplies, but the attendees ultimately decided not to recommend tougher restrictions. Explicitly rejecting the option of creating "an atrazine strike force to respond visibly and swiftly to problems," the group instead agreed on "a deliberate, incremental response."[68]

Just as the herbicide industry tried to coax the Agriculture Department into its fight with EPA, the formaldehyde industry in the early 1980s urged the Housing and Urban Development Department to set a standard on formaldehyde in pressed-wood products—even though HUD rarely gets involved in health and safety matters.[69]

Why would industry ask for regulation? It was the time-honored tradition of fighting fire with fire, as is clear in the fine print of the rule that HUD adopted in 1984. The regulation explicitly states that it overrides—or preempts—any other regulation written by the states. And it just so happens that the states of Connecticut, Massachusetts, Minnesota, and Wisconsin were at that time moving toward much tougher limits on formaldehyde, restrictions that would look at total formaldehyde in the air, not just the emissions from a single piece of particleboard, as HUD did.

Records of the Formaldehyde Institute show how frantically the industry worked in the early 1980s to push for the preemptive HUD standard. "Ed Canfield of Casey, Scott & Canfield raised the issue of what activity the Institute might take to enhance the ability [to] argue at some future date that state action has been preempted by federal action," the minutes of a 1980 meeting say. "The possibility of [Energy Department] or HUD preemption was discussed."[70] Two years later, the preemption issue was still being discussed. "Preemption is likely," the minutes of a 1982 meeting say, "assuming that HUD emphasizes that product standards are the only feasible standards and that HUD states that it intends its standards to preempt state law."[71]

The records also show how important the industry thought it was to push for "product" standards rather than limits based on the true level of formaldehyde in the air, which is known as an "ambient" standard. Georgia-Pacific lobbyist Kip Howlett noted in a memo dated November 14, 1980, that particleboard manufacturers did not know what standards it was possible to achieve at a "reasonable" cost. They would have to urge phasing in the standards. "He [William Groah, the technical director of the Hardwood Plywood Manufacturers Association] further observed that whether we like it or not, most regulatory agencies appear to be leaning toward ambient air standards. If they do, the National Particleboard Association and the Hardwood Plywood Manufacturers Association will recommend going to laboratory conditions and will attempt to show that product standards can predict ambient air levels."[72]

Godish of Ball State University calls the HUD regulation "the Georgia-Pacific standard" because it was based on what the biggest company in the industry already knew it could achieve. And at four-tenths of a part formaldehyde per million parts of air, it was four times greater than the concentration that typically causes irritation. Godish said that he has no objection to product standards instead of ambient standards, if they are set reasonably low—as, say, Sweden's are, at one-tenth of a part per million. However, the HUD regulation seemed designed to protect the industry from further regulation.

What's more, the HUD regulation protected the industry from something else: lawsuits. If corporations comply with all federal regulations in the manufacture of their products, the courts tend to view product liability suits more skeptically. And because HUD regulations required that buyers of mobile homes be warned about formaldehyde, the industry could argue that supposed victims knew what they were getting into, just as manufacturers of cigarettes have long argued.

In 1993, perchloroethylene manufacturers and dry-cleaning groups lined up behind a federal environmental standard that was more loophole than anything else. It was developed by the EPA's Office of Air and Radiation, which had been ordered by Congress in 1990 to set standards to control 189 known hazardous pollutants—which it had been unable to do for years. By the end of 1992, the EPA was supposed to have completed 40 of them but had finished work on only two. Under tremendous pressure to show results, the EPA approved the first such standard—for perc. From the industry's viewpoint, one of the rule's most attractive features was its exemption for small businesses. At least half of the nation's dry cleaners could ignore the regulations completely.[73]

Even the Center for Emissions Control, an arm of the halogenated solvents industry, felt that the rule was too weak. Steve Risotto, the center's director, says that the manufacturers' main criticism is that the

EPA did not even require the phasing out of the old-style "transfer" machines still used in a third of the industry, which require workers to manually move perc-drenched clothing from a washing machine into a dryer, exposing them in the process to high levels of perc. The chemical companies want to see dry cleaners invest in the newest technology, where washing and drying are done in the same machine with filters and condensers that further cut down on perc waste.

While the new technology means that dry cleaners use less perc, which means lower sales for industry, over the longer term, it ties a dry cleaner to a future with perc. Risotto openly admits the chemical industry's goal of preserving its market. "We'd like to see less perc used safely," he says, "than see none used at all." Greenpeace, Consumers Union of the United States, and other organizations argue that the newest machinery is not completely safe, because perc fumes still escape.

The Revolving Door

Manufacturers have many ways of building close relationships with key regulators. One of the little-known facts of life in Washington is that bureaucrats regularly receive travel and lodging expenses from the very corporations and trade associations that have the most to lose—or gain—from their decisions.

In 1989, acknowledging that paid travel for regulators raised ethical issues, Congress passed a law requiring that such travel be approved by agency managers and that public records be kept on all such paid travel. According to those records, EPA employees took at least 3,363 trips from March 1993 to March 1995 that were paid for—at a cost of more than $3 million—by corporations, universities, trade associations, labor unions, environmental organizations, and other nongovernmental sponsors. Four of the twelve corporations that this book examines closely—Ciba-Geigy, Dow, DuPont, and Monsanto—hosted EPA employees on at least 25 trips to their corporate headquarters and other locales.[74] Organizations closely connected with these corporations, however, financed many more trips. (See Table 4.1.)

Various agricultural organizations—including the American Crop Protection Association, the California Agricultural Leadership Association, the Florida Fresh Fruit and Vegetable Association, and the Iowa Corn Growers Association—paid for at least 41 trips by EPA employees

TABLE 4.1: AGENCY JUNKETS
EPA Employee Trips Paid for by Groups That Oppose
Stricter Regulation of Alachlor, Atrazine,
Formaldehyde, and Perchlororethylene,
1993–95

Sponsor	Number of Trips	Cost of Trips
American Chemical Society	31	$ 23,430
Dow Chemical Company	16	16,093
Society of Environmental Toxicology and Chemistry	12	14,162
Water Environment Federation	11	10,004
Toxicology Forum	10	11,038
Chemical Manufacturers Association	10	5,759
American College of Toxicology	9	5,595
Florida Fresh Fruit and Vegetable Association	9	7,179
Center for Emissions Control	7	4,600
Chemical Specialties Manufacturers Association	7	3,276
Iowa Corn Growers Association	6	5,075
American Crop Protection Association	6	3,175
California Agricultural Leadership Association	5	6,816
Environmental Information Association	5	3,505
Synthetic Organic Chemical Manufacturers Association	5	2,450
Ciba-Geigy Corporation	4	2,321
Chemical Producers and Distributors Association	3	3,665
E. I. du Pont de Nemours and Company	3	575
National Institute of Chemical Studies	3	2,160
Technical Association of the Pulp and Paper Industry	3	2,037
Vision 2000	3	1,935
TOTAL	168	$134,940

in the two-and-a-half-year study period. Dow topped all of the chemical companies, spending more than $16,000 for 16 trips to its corporate headquarters in Midland, Michigan, and to conferences in Fort Lauderdale, Florida. But to gain a true measure of Dow's largesse, one must also count the seven trips (costing at least $4,600) arranged by the Center for Emissions Control. "Vision 2000," a 1994 conference organized by the dry-cleaning and chemical industries to discuss perc's environmental and health problems, paid for another three trips.

The American Chemical Society sent EPA employees on trips at a rate of one a month from 1993 through 1995. The Chemical Manufacturers Association paid for 10 trips; the Chemical Specialties Manufacturers Association picked up the tab for seven. The Chemical Producers and Distributors Association, the industry-subsidized National Institute for Chemical Studies, and the Technical Association of the Pulp and Paper Industry each sent EPA officials on three trips in those two years.

Many officials of federal agencies view public service as an opportunity to do good. Some, however, view it merely as an opportunity.

The revolving door between government and industry is a fact of life and is perhaps the chemical industry's greatest weapon in its efforts to stifle regulation. The door spins in both directions. John Byington, the former chairman of the Consumer Product Safety Commission, has advised the formaldehyde industry on how to deal with his ex-employer. Linda Fisher, a former assistant EPA administrator for prevention, pesticides, and toxic substances, joined Monsanto in 1995 as its vice president for government affairs, heading the company's Washington office. Susan Vogt, one of Fisher's top deputies at the EPA, is now Georgia-Pacific's director of environmental policy, training, and regulatory affairs. Peter Voytek, the scientist who oversaw the EPA's first assessment of the risk of perchloroethylene in the early 1980s, is now the director of the industry's chief lobbying group, the Halogenated Solvents Industry Alliance. Peter Robertson, before becoming EPA Administrator Carol Browner's chief of staff in 1995, was the International Fabricare Institute's top Washington lobbyist. These examples illustrate how the revolving door tends to benefit the chemical industry—both going and coming.

The Center for Public Integrity was able to identify where half of the 40 EPA officials who left top-level jobs in toxics and pesticides during the past 15 years went to work. Of those, 18 went to work for chemical companies, their trade associations, or their lobbying firms; two went to work for environmental organizations.

Virtually all the major chemical manufacturers employ former toxics regulators, whether they served in Democratic or Republican administrations. Access is a bipartisan commodity. Don Clay, who headed the EPA's solid waste and toxics offices during the Bush admnistration, runs Don Clay Associates, Inc., a consulting firm whose clients include Dow Chemical Company and the Chemical Manufacturers Association. Steven Jellinek, the assistant EPA administrator for pesticides and toxic substances during the Carter administration, runs Jellinek, Schwartz & Connolly, Washington's top pesticide-lobbying firm. His winning formula: hire top talent from the EPA and tout their expertise and access. The firm's clients range from Dow and Monsanto to the National Cotton Council and the National Association of Wheat Growers.

Of the 344 lobbyists and lawyers that the Center identified as having worked from 1990 to 1995 for the chemical companies and trade associations that this book explores in depth, in fact, at least 136 previously worked for federal departments or agencies or in congressional offices. (See Table 4.2)

The impact is "immeasurable but significant," Rick Hind, the legislative director of Greenpeace's toxics campaign, says. "The many good and dedicated public servants who remain there at a career level are penalized for doing so, because they are not rewarded financially for it. And when a former assistant administrator comes back as an industry lobbyist, he has a psychological edge. It affects everything—how agency officials respect his schedule, his opinions, how they give him the benefit of the doubt."

James Conlon, who was deputy director of the EPA's Office of Pesticide Programs for eight years, calls turnover "a very large problem that has been exacerbated in recent years." Conlon, now retired, remembers helping to nearly double the size of the agency's pesticide program with a burst of hiring from 1978 to 1981. "We made a particular effort to bring some first-class people in, but virtually all of that generation is gone now," he says. "It's amazing how many of them skipped town, most of them to work at chemical companies. They were among the best and the brightest."

	Lawyers/ lobbyists	Industry employees	TOTAL	Comments
TABLE 4.2: THE REVOLVING DOOR **Government Officials Later Employed** **by the Chemical Industry, 1990—1995**				
Environmental Protection Agency	24	3	27	Includes 10 assistant administrators, four deputy assistant administrators, and two directors under the assistant administrator for prevention, pesticides and toxic substances
Agriculture Department	8	3	11	Includes two Agriculture secretaries, two assistant secretaries, and a deputy secretary
Congress	43	13	56	Includes four members of Congress
White House	6	4	10	Includes three special assistants to the president, a White House chief of staff, and a White House counsel
Food and Drug Administration	4	0	4	Includes a general counsel and an associate chief counsel
Other federal departments and agencies	20	8	28	Includes a chairman of the Consumer Product Safety Commission
TOTAL	105	31	136	—

NOTE: In cases where persons worked for more than one agency, priority was generally given in the descending order listed above. There was a single exception in the case of a former congressional aid who went on to become White House chief of staff.

5

Making Friends
in High Places

If there had been a crowbar handy, Mike Synar might have grabbed it the day he stood on the floor of the House of Representatives to implore his colleagues to repeal a law that required the government to buy back from manufacturers, wholesalers, and farmers the entire stock of any pesticide banned by the Environmental Protection Agency. It was a bold effort by the Oklahoma Democrat that June day in 1988 to pry apart the tightly fused bond that joins the chemical industry and Congress. "This may be the only vote we get this year," Synar (who was defeated for reelection in 1994 and died two years later) told his colleagues. "It may be the most important vote we have on pesticides."[1]

Synar had picked his target well. The so-called indemnification provision of federal pesticide law, which today survives in modified form, is widely regarded as a two-pronged boondoggle for industry: It discourages aggressive enforcement by the EPA while guaranteeing that manufacturers can profit even from products that are dangerous enough to ban. Amazingly, in some years the appropriations program has been so costly that the federal government has spent more money buying banned pesticides than it has spent regulating pesticides. In 1988, for instance, as Synar was voicing his objections, Congress was preparing to approve an appropriations bill for the 1989 EPA budget that allotted more money for

119

the purchase, storage, and disposal of banned products ($53 million) than for the registration, control, and testing of pesticides ($45 million).[2]

Outrageous or not, manufacturers were not about to give up indemnification without a fight—and neither were their friends in Congress. By the time Synar took the floor, Democrat E "Kika" de la Garza, then the chairman of the House Agriculture, Nutrition, and Forestry Committee, whose south Texas district was dotted with corn-fields and citrus groves, already had organized the opposition. Hearing of plans by Synar and Representative William Clinger, a Republican from Pennsylvania, to offer an indemnification repeal amendment to an unrelated appropriations bill, de la Garza arranged for the House Rules Committee to knock out Synar's amendment without a vote because, de la Garza argued, his committee had jurisdiction over any pesticide bill—an assignment that would be a death sentence for a repeal of indemnification.

When Synar appealed to the full House to overturn the Rules Committee's decision, de la Garza stood up and launched a counterattack. De la Garza argued, illogically, that indemnification—though hated by environmentalists—was actually an environmental program. The federal government ought to buy up banned pesticides, he said, "so that the farmer does not dump the suspended pesticide in the nearest creek, so that the retailer does not take it to the city dump."

An incredulous Synar retorted that no farmer had ever asked for indemnification but that manufacturers had reaped millions of dollars from the program. Chevron Corporation, for instance, was paid $12.8 million by the government when the weed-killer Silvex was banned in 1979, and other Silvex manufacturers and wholesalers were paid about $7 million more.[3]

The vote was a cliffhanger, but with heavy arm-twisting and a slew of vote switches after time had nominally expired, de la Garza won with a razor-thin margin, 209–206. The victory had a price: De la Garza had to promise the House that his committee would approve a pesticide bill later that year. But the bill his committee eventually wrote, and that President Reagan signed into law, stopped well short of repealing indemnification. The government would no longer have to compensate manufacturers for banned pesticides automatically, but Congress still could; if the manufac-turer of a banned pesticide was insolvent, the government would have to compensate farmers and distributors left holding supplies of the pesticide. The changes have slowed—but not stopped—indemnification.

The federal government is still compensating manufacturers for the 1986 banning of Dinoseb. The U.S. Treasury has paid out about $11 million so far, including $8.7 million to Uniroyal Chemical Company, Inc., Dinoseb's manufacturer.

The story of the weed-killer welfare bill that would not be killed is one of dozens of similar tales that could be pulled from the annals of the 25-year history of environmental law in the United States. This is true even though politicians want to be seen by the voters as guardians of public health. Polls consistently show solid public support for environmental and health protection, and one of the few mistakes that a top aide to Republican Newt Gingrich of Georgia admitted making in his first year as House Speaker was passing an overhaul of the streams and rivers protection law that became popularly known as the "Dirty Water Bill."[4]

Therefore, members of Congress, like de la Garza, who rise in opposition to environmental measures make valiant rhetorical efforts to show that they are working in the public's interest. But ordinary citizens do not take the chairman of the House Agriculture Committee to Palm Springs. They do not fly him to Las Vegas or New Orleans. A typical Texas couple worried about the effect of pesticides on their children would not be able to bankroll a congressional campaign, even if they could afford it, because of limits on what individuals can give a candidate.

The chemical industry can do, and has done, all these things. It brings the same types of pressure to bear on politicians as it does on decision makers in the federal bureaucracy—"grass-roots" organizations, reams of "science." But on Capitol Hill, lobbyists are permitted to use another powerful tool: money.

The corporations and trade associations with an interest in the four chemicals that are the focus of this book gave de la Garza at least $150,675 in campaign contributions from 1979 to 1995, according to reports filed with the Federal Election Commission.[5] But that's not all the Texan got. In just five years, from 1990 to 1994, de la Garza took 17 junkets and $7,000 in speaking fees from groups with an interest in the four chemicals.[6]

To appreciate the importance of the kind of money that the chemical industry can provide, one has to understand what campaign funds have come to mean in the political process. By the 1994 elections, the average senator had to raise more than $14,500 a week, every week, during his or her entire six-year Senate term to wage a winning campaign—40 percent more than just eight years earlier. The average cost of a successful House campaign—about $73,000 in the post-Watergate election of 1974—had risen nearly sevenfold, to more than $500,000 by 1994, or nearly $5,000 in weekly fund-raising over a two-year term.[7]

The average citizen cannot—and under law, must not—contribute more than $1,000 per election to a congressional campaign, a fraction of any candidate's weekly fund-raising needs. But corporate giants can spend as much as they want through the deft use of political action committees and a variety of other avenues to channel money and influence to the only people who have the power to put truly effective restrictions on their products.

One of the goals of the post-Watergate campaign finance reforms was to stop direct giving to candidates by corporations. Instead, companies would set up political action committees to collect voluntary contributions from employees, with all contributions and expenditures on the public record. But in many ways, as Charles Lewis points out in the Center for Public Integrity's 1996 book *The Buying of the President*, these reforms served to sanction and legitimize the influence of money in politics.

Companies may portray their PACs as representing a broad cross-section of their workforce, as with the G-P Employees Fund of Georgia-Pacific Corporation. But of the company's 50,000 employees, only 10 had given to the PAC in 1995 and the first half of 1996, and eight of them were in the highest ranks of corporate management: Alston D. Correll, the chairman and chief executive officer; Harvey C. Freuhauf Jr., a director; Donald L. Glass, James F. Kelley, John F. McGovern, and Lee M. Thomas, all senior vice presidents; Davis K. Mortensen, an executive vice president; and John M. Turner, the vice president for government affairs.

What does a corporation like Georgia-Pacific hope to get for its money?

The company's lead environmental lobbyist, Kip Howlett, probably expressed it most clearly in the early 1980s. With formaldehyde under fire by the federal agencies, he talked about the importance of gaining influence in Congress in discussions with his fellow Formaldehyde Institute members. He spoke of "playing the saccharin ace trump card"—a reference, they knew, to the law Congress had passed in November 1977, seven months after the Food and Drug Administration proposed to ban the sale and marketing of saccharin based on numerous scientific studies that linked the sweetener to malignant bladder tumors in rats.[8] Congress put into effect an 18-month moratorium that barred the FDA from using any funds to enforce its saccharin ban; Congress has renewed that "temporary" measure repeatedly, and it remains in place today.[9]

This Saccharin Study and Labeling Act remains a symbol of the laws that have emerged from Congress over the years to deal with consumer hazards: rife with language that purports to protect public health—and riddled with wide-open loopholes to protect industry.

The Avenues of Influence

In theory, corporations are limited in the amount of money their PACs can give to candidates for federal office—$5,000 per election. In reality, those dollars multiply like proverbial loaves and fishes through the miracle of the campaign finance system. This becomes clear by looking at the political giving patterns of the 12 corporations that manufacture the four chemicals this book examines in depth. The group includes the giants of the industry (Ciba-Geigy, Dow, DuPont, Hoechst Celanese, ICI, and Monsanto) as well as smaller manufacturers (PPG Industries and Vulcan), and more diversified companies (Borden, Georgia-Pacific, Occidental, and Weyerhaeuser).

Their corporate PACs gave congressional candidates more than $7 million from 1979 through 1995. (See Table 5.1.) Add all the contributions from PACs operated by associations, institutes, and other organizations affiliated with those corporations, and the total rises to more than $20 million.

Then there are the avenues of influence that can be opened up by someone like Buck Mickel, a South Carolina industrialist, a member of Monsanto's board of directors, and a leading fund-raiser for Republican candidates in his state. The benefits of the mutual backscratching between the chemical industry and Congress are neatly symbolized by an exchange of letters in 1985 between Mickel, Republican Senator Strom Thurmond, and the EPA. The letters were obtained from the agency through the Freedom of Information Act.

In his July 31, 1985, note to Thurmond, Mickel described Monsanto's concerns about the agency's review of alachlor. He complained that EPA officials were reluctant to meet with representatives of Monsanto, enclosed a "private briefing paper" summarizing the company's

TABLE 5.1: CAMPAIGN CASH
Contributions to Congressional Candidates from Political Action
Committees and Employees of 12 Chemical Companies

Company	PAC Contributions, 1979–95	Contributions by Employees, 1986–94
Dow Chemical Company	$1,454,615	$ 555,628
Weyerhaeuser Company	877,973	152,435
Georgia-Pacific Corporation	867,067	19,500
Hoechst Celanese Corporation	806,175	39,700
Ciba-Geigy Corporation	765,643	24,670
E. I. du Pont De Nemours and Company	675,535	119,250
Monsanto Company	663,792	126,167
Occidental Petroleum Corporation*	385,782	166,904
PPG Industries	322,540	44,000
Imperial Chemicals Industries, PLC	249,830	22,498
Vulcan Materials Company	214,675	98,050
Borden, Inc.	67,800	4,649
TOTAL	$7,351,427	$1,373,451

*Figures through 1991; Occidental stopped producing perchloroethylene in 1992.

arguments against stricter regulation, and urged Thurmond to bypass the bureaucracy and directly contact EPA Administrator Lee Thomas, who had been appointed by President Reagan.

"It is essential that the message get to Lee and that he understand what is happening for only he can head this off," Mickel's letter concluded. "I am personally in your debt for helping on this and deeply appreciate your agreeing to do this for me. Warm regards to you, Nancy, and your family. My best, Buck."

Today, Mickel (who retired from Monsanto's board of directors in 1996) says that he cannot remember sending the alachlor letter but acknowledges that he sometimes discussed business matters with Thurmond. "Sure I did, he's my senator."

Coming from a man with Mickel's track record as a fund-raiser, a request for a political favor can get quick results. Mickel, who at the time was the president of Fluor Corporation (the building-products giant), has long been a major player in GOP politics in South Carolina. And Mickel had already shown Thurmond that he was willing to put his money where his mouth is: Federal Election Commission records show that he personally contributed $2,000 to Thurmond's 1984 reelection campaign, while Fluor's PAC kicked in another $7,000—more than it gave to any of the other 37 senators it supported in 1983 and 1984.[10]

Thurmond swung into action with a polite but pointed letter to Thomas dated August 15, 1985. "Apparently there has been a breakdown in communications, and I am confident that, with your intervention and participation, this matter can be fairly reconciled," he wrote. Thomas answered on October 3, 1985, with a letter assuring Thurmond that "there has been some misunderstanding" and promising him that EPA officials would meet with Monsanto to talk about alachlor. Eventually, the EPA would do much more than that: The agency would abandon its original plan to ban alachlor and instead adopt a package of watered-down restrictions that Monsanto supported.

Over-the-counter political money is only a fraction of the currency that flows between the nation's chemical companies and Capitol Hill lawmakers.

For instance, 17 members of Congress who sat on the six committees with primary jurisdiction over chemical regulation from 1990 to 1994 owned stock in the companies that produce alachlor, atrazine, formaldehyde, or perchloroethylene, and they received a total of at least $26,733 in dividends and capital gains from those investments. The minimum collective value of their holdings—calculated from their financial disclosure forms—was $211,017. This situation, though perfectly legal, literally gives the lawmakers a stake in the ups and downs of a chemical company's fortunes—and a ban, new regulation, or other change in the status quo could cost them money. In 1996, these stockholding lawmakers included three current and former committee chairs, the people most responsible for deciding the agenda for the health and safety laws Congress considers each year: Republican Thomas Bliley of Virginia, the chairman of the House Commerce Committee; Democrat John Dingell of Michigan, Bliley's predecessor as the committee's chairman; and Republican Nancy Kassebaum of Kansas, the chairwoman of the Senate Labor and Human Resources Committee.[11]

Another way to develop friends in high places is to wine, dine, and fly them. Public records show that chemical companies frequently host influential politicians on trips to lush resorts. From 1990 to 1994, the chemical companies and trade associations with an interest in the four chemicals examined in this book gave members of Congress at least 214 expense-paid trips, including 95 members of six key committees. (See Table 5.2.)

The National Agricultural Chemical Association, for example, paid for 38 trips by Capitol Hill lawmakers from 1990 to 1994. It flew Representative Pat Roberts, a Republican from Kansas, to Rio de Janiero in 1991; the Kansas Republican also took three agribusiness-financed trips to Florida over a four-year period. Roberts received $123,725 in chemical industry PAC funds from 1979 to 1995 and $6,000 in speaking fees from 1990 to 1994. He proved to be a good investment: In 1995 he became the chairman of the House Agriculture Committee, where he was a staunch industry ally.[12]

Another industry favorite, Representative Charles Hatcher, a Democrat from Georgia, got junkets to Albuquerque, Orlando, San Diego, and Naples, Florida—and $11,750 in speaking fees in 1990 and 1991—from pesticide manufacturers and other agribusiness interests.[13] The following year, just before he was defeated for reelection, Hatcher led an unsuccessful fight for a proposal that would have barred local governments from passing their own pesticide regulations. Such a change was needed, he argued, because the current system is "unduly burdensome" to chemical companies.[14]

Representative Charles Stenholm, a Democrat from Texas, has been another frequent junketer. From 1990 to 1994, he took 25 trips and collected $23,950 in speaking fees from companies or associations with an interest in the four chemicals examined in this book. A senior member of the Agriculture Committee and a cotton farmer with extensive land holdings in his home state, Stenholm took in at least $121,897 in PAC contributions from those interests from 1979 to 1995. He, too, has been a loyal industry ally. In 1992, a year in which he took six industry-financed trips and collected $3,200 in speaking fees from those interests, Stenholm pushed unsuccessfully for an amendment that would have given manufacturers the right to sue the EPA for damages if the agency banned a pesticide capriciously.[15]

TABLE 5.2: CONGRESSIONAL JUNKETS
Trips by Members of Six Key Congressional Committees* Paid for by Groups That Oppose Stricter Regulation of Alachlor, Atrazine, Formaldehyde, and Perchloroethylene, 1990–94

Sponsor	Number of Trips
American Crop Protection Association**	40
Grocery Manufacturers of America	11
American Farm Bureau Federation	6
Farm Credit Council	6
National Cotton Council	6
United Fresh Fruit and Vegetable Association	6
American Association of Crop Insurers	5
Monsanto Company	5
Southwestern Peanut Shellers	4
Cotton Warehouse of America	4

*House Committees: Agriculture; Commerce; Economic and Educational Opportunities. Senate Committees: Agriculture, Nutrition and Forestry; Environment and Public Works; Labor and Human Resources Committees.
**The American Crop Protection Association was known as the National Agricultural Chemicals Association until September 20, 1994.

Bonding sessions with politicians are important for the industry, which gains the chance to arm its friends in Congress with effective (though often misleading) arguments for use in drafting legislation. One well-traveled piece of hyperbole—that a person would need to drink 38 bathtubs of water per day to experience a risk from atrazine in drinking water—even made it into a "Dear Colleague" letter that Republican lawmakers circulated in an effort to win support for curbing EPA's regulatory authority. EPA Administrator Carol Browner denounced the story as false, pointing out that drinking just two liters of water per day results in a 1 in 100,000 cancer risk—equivalent to 2,600 additional cancer cases per year nationwide.[16]

The Value of Access

A "dry-cleaning breakfast" may not sound like the most appetizing morning meal, but it was an important event for members of that industry who gathered in a Capitol Hill dining room in October 1994. They knew what was on the menu: access.

The International Fabricare Institute, financed by Dow as well as by small dry cleaners, also had a message to deliver that day. It handed to members of Congress who attended the breakfast a three-page briefing paper on the EPA's recent efforts to regulate the dry-cleaning business. It was titled "Fast Facts," a term that seemed only 50 percent accurate to several EPA staffers who, unbeknownst to the institute, had been invited to the breakfast by a group of dry cleaners from California.[17] Knowing their agency's glacial progress on developing tougher dry-cleaning regulations, they sat stunned to read that EPA's "recent initiatives...could threaten the existence of the dry-cleaning industry as we know it."

The IFI said that the EPA was considering the "possible elimination of perchloroethylene in favor of other untried non-solvent-based cleaning methods." In fact, the EPA's Office of Pollution Prevention and Toxic Substances had done nothing more than test and analyze the cost and effectiveness of alternatives to perchloroethylene—much to the dismay of others in the agency who felt that the cancer risks demanded urgent action. The only additional step that the EPA had contemplated, with great hesitation, was the publication of its analysis of risks and benefits of various cleaning technologies. That report would spell out those cancer risks clearly to the public for the first time, and the industry groups had no intention of letting it go forward.

This work, the IFI argued, should stop. Environmental issues in the dry-cleaning industry already had been addressed, it argued, by an EPA regulation in 1992 that required "all new and most existing dry-cleaning machines" to be equipped with "refrigerated condensers," the latest technology to capture fumes. In effect, the cool condenser recirculates the perc-contaminated steam in the machine instead of allowing it to escape. What the IFI did not mention was that the Clean Air Act rule contained an exemption for small businesses—virtually the entire dry-cleaning industry. And there were serious doubts that the refrigerated-condenser technology could truly control the perc risk to people who lived or worked in buildings with dry-cleaning establishments.

Within weeks of the October 1994 breakfast, four members of Congress—Martin Frost of Texas, Jon Christensen of Nebraska, Jim Ramstad of Minnesota, and Rob Portman of Ohio—requested a meeting with top officials of the EPA to talk about the agency's work on the dry-cleaning risks and benefits report. The record of the meeting shows how effectively the chemical industry has obscured the way lawmakers lobby for big business in the name of small business.

Lawmakers do not want to be seen as water carriers for corporate America; they would rather be seen as doing favors for the little guy. Consequently, the chemical industry typically turns to farmers, shop-owners, homebuilders, and the like to do their bidding. The EPA's planned report would tell for the first time the stark numbers about the cancer risks of dry-cleaning chemicals, as well as the economic advan-tages of the alternatives—information that presumably would benefit small-business owners more than anyone. But the four lawmakers who met with EPA officials invoked mom-and-pop shop owners—their "constituents"—in their argument about "science" and "risk" that other-wise was a page right out of the chemical industry's lobbying book.

"As you probably know, our constituents expressed their opposition to releasing part of the [Cleaning Technologies and Solvents Assessment] report," they said in a follow-up letter after the meeting. "We know you are committed to the use of good science in EPA decision-making. Based on what we are hearing from our constituents, however, we are concerned that your staff may not be following expressed policy. If flawed numbers are released, it could do harm to the dry-cleaning industry and to my constituents. We are sure that you understand the gravity of this situation."[18]

Two months before the dry-cleaning breakfast, one of the four lawmakers, Republican Jon Christensen of Nebraska, received a $1,000 campaign contribution from the PAC operated by Dow, a manufacturer of perchloroethylene.[19] Christensen and another attendee, Martin Frost, a Democrat from Texas, got lots of dry-cleaning money in the months following the breakfast. In April 1995, Frost received $2,000 from a couple who owned a dry-cleaning establishment in Irving, Texas. Christensen got a total of $4,000 from a couple in Omaha who listed themselves as "Martinizers," a class of franchisees who have been particularly active on the lobbying front.[20]

Whether the chemical at issue is perc or a pesticide, the threat that industry will call in its chits and get Congress to intervene in the regulatory process, or even cut the EPA's budget, is an ever-present concern at the EPA. "The EPA doesn't doubt that with three phone calls Monsanto or Ciba-Geigy can flood them with letters from the appropriations committees threatening to screw them if they don't fly right," says Charles Benbrook, a onetime Capitol Hill staffer and former director of the Board on Agriculture at the National Academy of Sciences.

When Monsanto was facing the possible banning of alachlor, for instance, it had politicians bombard the EPA with letters. In 1985, Governor Terry Branstad of Iowa wrote a letter to the EPA in which he pointed out that Monsanto's alachlor factory provided 600 manufacturing jobs in Muscatine, Iowa, and that the chemical was a "key element to many farmers' crop management programs."[21] Representative Jim Leach, a Republican from Iowa, wrote a similar letter, which began "The Monsanto Company has contacted me. . . ."[22]

Why do agency officials worry about such letters from Congress? Because lawmakers hold their purse strings, and their futures, in their hands.

The Consumer Product Safety Commission learned that the hard way when it proposed a ban on urea-formaldehyde foam insulation just before the 1980 elections. After Republicans gained control of the Senate following Ronald Reagan's landslide victory, lawyer-lobbyist Mary Martha McNamara assured members of the Formaldehyde Institute, "The climate in Congress will be different because of the recent national elections."[23]

Less than three months into the new congressional session, E. Josh Lanier, the executive director of the National Insulation Certification Institute, told a Senate committee that the CPSC was destroying small businesses across the United States. "The industry I represent here this morning is mad as hell and is disappointed and hurt by the way it has been treated by its own government," he said. "The urea-formaldehyde foam insulation industry has been bullied, abused, smeared in the press, maligned through special consumer alert bulletins and telephone hotlines, and, finally, crippled and brought to its knees by the Consumer Product Safety Commission. As we speak here this morning, a once promising and emerging industry in the energy conservation field is on the verge of bankruptcy. Without immediate corrective action, an entire industry—one made up exclusively of small businesses—will become a victim of regulatory abuse of the most sordid level."[24]

By June, the Formaldehyde Institute had made considerable headway in persuading members of Congress to slash the CPSC's appropriation and to take away its power to oversee the problems that formaldehyde insulation, wood paneling, and flooring products had caused.

"The final bill contains a fair amount of everything the Institute worked for," S. John Byington, another lawyer-lobbyist for the industry, said at a meeting of the Formaldehyde Institute.[25] The irony was that Byington had been the chairman of the CPSC; now he was on Capitol Hill urging Congress to take money and power away from his old agency. The Formaldehyde Institute's minutes of that period read as if *its* members, instead of members of Congress, were writing the law: "It was decided that efforts should go forward to seek to deprive CPSC of all jurisdiction over building products, without specifying any one agency [where] authority over building products should be vested."[26]

The Formaldehyde Institute did not get everything it wanted. It had hoped to change the structure of the CPSC to give it a single administrator appointed by President Reagan, instead of having three commissioners with staggered terms that overlapped administrations. Since the establishment of the Interstate Commerce Commission at the turn of the century, Congress had set up many regulatory agencies as commissions to avoid partisanship in regulation. As the formaldehyde industry saw it, the idea was working too well at the CPSC, the youngest of the federal commissions.

Congress did not take away the CPSC's status as an independent commission, nor did it deprive the agency of authority over building products. But it slashed its annual budget from $42.1 million to $32.1 million, and its staff was cut 28 percent, from 889 to 636. The Consumer Product Safety Commission never took on the formaldehyde industry again. Instead it worked with the industry to "monitor," but never regulate, the fumes from wood products, which it already knew was a worse problem than the fumes from the foam insulation it earlier had sought to ban.

The Art of the Loophole

The most obvious way to assess the chemical industry's success in making friends in high places is to look at the peculiar shape of laws on consumer health and safety. Since the original Earth Day in 1970, Congress has passed enough environmental laws to ensure that the United States has more protections of air, water, land, and health—in theory—than any other nation in the world. In practice, however, each of those laws is full of loopholes in just the right places for industry.

In its 1990 overhaul of the Clean Air Act, for example, Congress seemed to be acting for the first time to correct the EPA's dismal record of inaction on hazardous air pollutants. Among other things, it created a list of 189 well-known hazardous pollutants and mandated that the agency, for the first time, set national control standards for them.

To know what this portion of the law really meant, however, required a close reading of the voluminous fine print. The new law was chock-full of provisions that would be utterly incomprehensible to the average citizen but amount to escape clauses for the chemical industry, whose lobbyists were there each day to shepherd the bill to its final form. Owing largely to the sophisticated language the lobbyists developed over years of dealings with the regulators and the courts, the Clean Air Act swelled from the original 40-page statute passed in 1970 to the 700 pages of amendments passed in 1990.

First of all, the revamped law did not take the simple course of setting limits for different toxic pollutants. Instead, "major" polluters would be required to use the "maximum available control technology," a slippery definition on which the EPA, industry, and environmentalists could and would battle for years for each of the 189 pollutants.[27] In addition, there would be different, less difficult standards for smaller polluters, opaquely dubbed "area sources" by the law, and including as big a segment of each industry as the lobbyists could negotiate with the EPA—again, for each pollutant.

Environmentalists were concerned about the law, born out of the disjointed way that the United States had chosen to protect the environment. Laws covering air, water, and land are separate and have different levels of stringency. Now, with a newly strengthened Clean Air Act, environmentalists worried that industry might simply meet the requirements by dumping toxics on land or in sewers instead of burning them. They persuaded Congress to require that the EPA, when setting the standards for each pollutant, take into consideration "any non-air quality health and environmental impacts."

Dry-cleaning establishments, which had been reducing perc exhaust from machines with filters that they would later dump in the trash, realized immediately that the clause could have profound implications for them.

On the Senate floor, Republican Steve Symms of Idaho added an amendment to the law to spell out that for "area sources"—the little polluters—EPA should not analyze the impacts of its regulations on anything but the outdoor, "ambient" air. The language did not mention dry cleaning, though the industry's stake in the amendment was obvious to those who had been lobbied by the chemical companies and the cleaners.

"A number of small businesses in our state, many of them representing small-town dry cleaners, have contacted me regarding this amendment," said Senator James McClure, a Republican from Idaho, in support of the amendment. "To them, this clarification is absolutely vital." Senator Daniel Patrick Moynihan, a Democrat from New York, agreed. "It was not the committee's intent," he said, "to reshape the dry-cleaning industry."

These words would haunt the EPA before the ink was dry on the air regulation that it negotiated with the dry-cleaning industry. New York health officials had collected some frightening data on the level of perc fumes in apartment buildings with dry-cleaning establishments, and data had come in from California suggesting that groundwater contamination from dry-cleaning establishments was pervasive.

In 1993, when the EPA announced a meeting to explore whether it should look at the problems of indoor air pollution and groundwater contamination from perc, the dry-cleaning industry was prepared with the pages from the *Congressional Record* that had the speeches of Symms, McClure, and Moynihan.

"The legislative history of the Clean Air Act Amendments of 1990 makes clear that Congress intended that no residual risk analysis be conducted for area dry cleaners," William Fisher, a lobbyist for the International Fabricare Institute, said in a letter to the EPA. "For the agency to do otherwise would be to flout the intent of Congress."[28]

There is no question that the intent of Congress in this portion of the Clean Air Act was shaped by the intent of industry. "The halogenated solvent group [representing manufacturers of chlorine-based chemicals like perc] was an early supporter of technology-based standards," says Steve Risotto, the director of the industry-financed Center for Emissions Control. (Instead of setting limits on the amount of chemical that can be released into the air, the EPA by law must set standards on the technology that should be used to control emissions.) "I don't want to say we were the father of the idea. It's clear you needed a standard you can live with. You can sort of watch industries—the first time something is presented to them, immediately, they're going to fight it. But more and more, you realize you're better off being a part of it, rather than fighting it."

It is as good an example as any of what journalist William Greider has called "deep lobbying"—setting limits on the public debate at a very fundamental level. Environmental activists, who had seen EPA make only the slightest headway in setting limits on toxins, believed that the industry's technology-based standards would finally put protections in place. But now, because the EPA must debate with industry on what technology to use for every one of the 189 toxic air pollutants that Congress identified in 1990, progress is as slow as ever.

Water-Down Lobbying

Deep lobbying by the chemical industry has been a fact of life in Washington as far back as 1947, the year the Congress passed the Federal Insecticide, Fungicide, and Rodenticide Act. Written with extensive input from chemical manufacturers and agribusiness, FIFRA was designed not to ensure that pesticides were safe but that they were lethal—that they would effectively kill insects, fungi, and weeds. While the law required some basic testing for acute health effects such as skin rashes and dizziness, the tests were oriented toward protecting farmers, not consumers. The EPA took over the enforcement of FIFRA in 1970, and two years later Congress changed the law to require testing for long-term health effects of all pesticides—including the approximately 50,000 already on the market.

But Congress, fed a steady diet of industry dollars, made sure that the law's fundamental anti-consumer slant did not change with the 1972 amendments. Other landmark environmental laws of the early 1970s, such as the Clean Air Act and the Clean Water Act, followed a straightforward formula: They set health-based standards and forced polluters to comply. Those laws mostly came out of the environment and public works committees of the House and Senate, but FIFRA is the domain of the House and Senate agriculture committees, whose top priority has always been the protection of agribusiness. That is why the 1972 amendments to FIFRA required the EPA to allow the sale of a pesticide—no matter how dangerous—unless it posed an "unreasonable risk to the environment, taking into account the economic, social, and environmental costs and benefits of the use of any pesticide."[29]

In its subsequent tinkering with FIFRA, including the most recent revisions in 1988 and 1996, Congress has not touched the cost-benefit language. Instead of setting an unambiguous, measurable safety standard that a pesticide manufacturer must meet to avoid a ban, FIFRA stacks the deck in favor of manufacturers. Before the EPA can ban a pesticide, it has to use industry-generated data to make subjective judgments about the chemical's "benefits," then somehow determine whether the pesticide poses an "unreasonable risk" in light of those benefits. With so many decisions to make, the EPA is particularly vulnerable to political influence, political pressure, and slanted data.

Another loophole in FIFRA forces the EPA to keep information about the so-called inert ingredients of pesticides secret. Inert ingredients often compose more than 95 percent, by volume, of a pesticide product, yet they are largely exempt from public disclosure because under FIFRA they are considered trade secrets. A pesticide product's ingredients are considered inert if they do not help kill the particular insects or weeds targeted by the product's active ingredients. However, inerts are anything but benign to the people who use them; they sometimes include toxic chemicals that are as harmful as the active ingredients.

The industry is so deeply involved in writing legislation that lawmakers have sometimes been embarrassed to learn that they do not know much about the bills they are sponsoring.

In 1984, for instance, Representative Dan Glickman of Kansas, whom President Clinton named in 1995 to be Agriculture Secretary, sponsored a relatively noncontroversial bill that would have given pesticide manufacturers greater patent protection. What he did not realize until sharp-eyed environmental lobbyists pointed it out to him was that the bill, which had been drafted by the National Agricultural Chemicals Association, also included a clause that would have overturned a recent Supreme Court ruling requiring manufacturers to make public the health and safety data they submitted to the government. An embarrassed Glickman said that the clause had been added without his authorization and would be removed, but the disclosure of the trade association's effort to slip the provision into the bill doomed it. The patent provision would not become law for another four years.[30]

In the 1990s, pesticide manufacturers fought long and hard to overturn the rule they hated most: the Delaney Clause. Ultimately, they succeeded. The law, written in 1958 by Representative James Delaney, a Democrat from New York, stated that processed foods could not contain any residues of pesticides that have been shown to cause cancer in laboratory animals. When the EPA tried in 1987 to adopt a weaker rule that would have allowed pesticide residues that pose "negligible" cancer risks, environmental organizations sued, and in 1992 they won a federal appeals court ruling that ordered the agency to enforce the Delaney Clause's zero-risk standard by 1997. That would have meant big changes for as many as 77 pesticides that the EPA says cause cancer in lab animals and are found in food. Few if any of those 77 pesticides would have been completely banned, but the EPA would have been forced to significantly reduce the number of crops that can be treated with them. Monsanto, for example, would have lost one of the biggest markets for alachlor: soybean farmers. Alachlor also would have been banned from peanuts and sunflower seed, as would have atrazine's use on sugarcane.[31]

But none of that will happen now because Congress legislated the Delaney Clause out of existence when it reauthorized FIFRA in 1996. It was the culmination of a four-year lobbying compaign by pesticide manufacturers, which began pouring PAC money into Capitol Hill after the Supreme Court's decision in 1992. A study by the Environmental Working Group found that PAC contributions from pesticide manufacturers during the first 18 months of the 103rd Congress topped $3.1 million—almost twice what they were during the first 18 months of the two previous Congresses.[32]

The Delaney Clause had many critics, even among environmental organizations. Illogically, it covered only processed foods, though raw fruits and vegetables contain pesticide residues that are just as high. (The EPA decides how much residue to allow on raw foods on a case-by-case basis.) In addition, the clause was written when testing methods were primitive and pesticides could be detected only in relatively high concentrations. Modern testing methods routinely detect residue levels as small as one part pesticide per billion parts food.

In 1993, the Clinton administration asked Congress to rewrite FIFRA and set a consistent rule that would apply for all foods, raw and processed. The administration's proposal: A pesticide should be used on a particular type of food only if residue tests show it present at levels that would pose less than a one-in-a-million risk of developing cancer to a person who ate a typical amount of that food for a lifetime.

The chemical industry had other ideas. It pressed Congress to drop the Delaney Clause and substitute a new standard that would have forced the EPA, in setting the residue limits, to consider their financial impact on farmers. The proposal stalled, even after the Republicans won control of Congress in 1994. So the industry adopted a back-door approach, convincing the House of Representatives in 1995 to approve a rider to the EPA appropriations bill that would have forbidden the agency from enforcing the Delaney Clause. The move failed in the Senate, however, and when public opinion turned against the Republican Congress in 1996, the industry relented. With the EPA's planned bans looming in 1997, chemical manufacturers agreed to accept the Clinton administration's one-in-a-million proposal—a compromise favored by many, but not all, of the major Washington-based environmental organizations. It is uncertain how many pesticides will be affected under the new standard, but the total is likely to be far less than the 77 that would have been subject to further restrictions under the Delaney Clause.

The Cost-Benefit Conundrum

When Capitol Hill lawmakers began to look at environmental problems in the late 1960s, they did not try to place limits on the manufacture or sale of toxic chemicals and products. Instead, they focused on the obvious—and overwhelming—problems of air and water pollution.

The deaths of thousands of workers from asbestos-related lung disease ultimately prompted Congress to set out to attack the problem of hazardous substances in the late 1970s. Like so many such ideas, however, the original proposal was chipped into a new shape by the skillful chisels of the chemical lobby. First, eight products were carved out of the bill on the rationale that there were other laws to take care of any problems they posed: pesticides, tobacco, nuclear material, firearms and ammunition, food, food additives, drugs, and cosmetics.

More significant, the 1976 Toxics Substances Control Act contained a tiny loophole that would wear ever thinner at the edges each time the federal government moved against a chemical. Eventually, the loophole would grow big enough to swallow and kill every genuine attempt to control the sale of harmful chemicals. The law would give the EPA tremendous power to control the sale, manufacture, and distribution of chemicals. First, however, the agency would have to weigh the costs to industry against the benefits to the public for every proposed action against toxic substances and prove that it had chosen to use the power in a way that was the "least burdensome" to industry.

The chemical industry has the resources to produce as many instant economic and scientific analyses on burdens and options as there are consultant-hours in the day. It has used those two words to delay—and sometimes thwart—regulatory action. Although there are some 70,000 chemicals in commerce, the EPA has been able to muster proof to overcome the cost-benefit test for only nine chemicals in the law's 20-year history.

In the decision that many would view as the death knell for the Toxic Substances Control Act, a federal court ruled in October 1991 that the EPA had not met the cost-benefit test for the crown jewel of its toxic regulations: its 1989 ban on most products containing asbestos. Not only was asbestos the substance that had prompted Congress to pass the toxics law in the first place, but the agency had spent 10 years gathering ever more compelling evidence of its dangers. Still, the chemical industry argued successfully that under the language in the law that its own lawyer-lobbyists had written, the EPA must consider the costs and benefits of each regulatory option—starting with labeling, working up through a variety of partial bans, and, finally, to the option of a full ban.[33]

After the decision, the EPA dropped its efforts to ban asbestos products. Moreover, officials of the agency told the General Accounting Office, when it was investigating why the agency had failed to take more actions against toxic chemicals, that it was unlikely the EPA would ever again devote its time and money to attempting to ban a product.[34]

"A lot of people felt that if you can't regulate asbestos under the Toxic Substances Control Act, you can't regulate anything," says George Semanyiuk, the EPA's formaldehyde project manager. "When the asbestos ban was overturned, a lot of people jumped to the gun of believing that was the death penalty of Section 6"—the part of the law that gives the EPA what seemed to be broad power to ban dangerous products. He holds out the hope that the law still could work if the agency concentrated on milder actions short of outright bans on products, although he admitted that he had spent a decade on largely unsuccessful attempts to obtain a voluntary industry program to control and test wood products that contain formaldehyde—products that he believes are widely used and extremely dangerous.

"I've seen the hope of the institution building up, and in the later days, it's quite sad, when you look back," he says. "It [the Toxic Substances Control Act] took so long to pass. Now, we're deluged with chemicals and spend millions and millions of dollars worth of resources on economic risk-benefit analysis." All this time and money, he says, and yet no closer to the goal of protecting public health.

One of the flashiest attempts to dress up cost-benefit analysis was in the "Contract With America," the manifesto that helped Republicans gain control of Congress in 1995 for the first time in 40 years. The GOP leadership's abortive effort to curb environmental regulation with cost-benefit analysis in the 104th Congress was called the Job Creation and Wage Enhancement Act.

House Speaker Newt Gingrich was only taking the next logical step in a movement that had begun in full force in the early 1980s. Even though it was a decade before the asbestos decision, the chemical industry had already clearly realized the value of the cost-benefit language it had placed in the Toxic Substances Control Act. No one was more willing to take up the cause than President Reagan, who had vowed to get government off the backs of Americans and who had mused during his campaign that even trees cause pollution. After Reagan's election, the chemical industry worked in earnest to have the idea extended to all health, safety, and environmental regulations. "Some *new* members of Congress are considering an initiative for developing broad guidelines on approaching such a regulatory issue as the Formaldehyde Institute faces," lawyer-lobbyist Mary Martha McNamara told members of the Formaldehyde Institute, according to the minutes of a meeting. The minutes went on to say, "She believes it will find a more favorable reception, especially in the Senate," where the GOP had taken control on the coattails of Reagan's victory.[35]

Kip Howlett, Georgia-Pacific's lobbyist, agreed, telling the group that he hoped formaldehyde would be a "case history" for something he called "regulatory relief." Soon, regulatory relief—dressed up, to be sure, in language tailored to appeal to the average Joe and Jane—would become an article of faith within the Reagan administration.

In his first month in office, Reagan signed an executive order that every new government program to protect health, safety, or the environment would be scrutinized by the Office of Management and Budget to see whether its intended benefits outweighed the likely costs to business. A little more than a year later, the Republican-controlled Senate passed a bill based on the idea without a single "no" vote.

However, similar legislation did not pass the Democratic-controlled House, and the Reagan administration was soon in so much trouble for cutting environmental sweetheart deals with industry that it abandoned the broader reform package.* Nonetheless, the cost-benefit calculus has never fully lost its luster; it has been polished up for use by politicians of both parties again and again.

Even though President Reagan could not get Congress to change the law, he made sure that the Office of Management and Budget scrutinized every major regulation for its costs and benefits, slowing down the already laggard pace of agency decision making. President George Bush assigned Vice President Dan Quayle to do the same job, and the White House's Council on Competitiveness blocked a variety of regulations meant to protect public health—the EPA's attempt to ban the incineration of lead batteries, for example. President Clinton closed down the Council on Competitiveness with a flourish when he took office, but he issued an executive order continuing the policy of weighing costs and benefits with every regulatory decision.

The 104th Congress took office in 1995 with a plan to take cost-benefit analysis much further, to do, in effect, what Reagan had tried and failed to accomplish. The plan in the Contract With America called for many different levels of study and analysis before any new regulation could be proposed.

Sara Schotland, a longtime attorney for the formaldehyde industry, says that one of the ideas in the GOP plan—peer review of any agency decision by outside panels of scientists—is an important one. Too often, she says, the industry encounters bureaucrats who have "married" an issue; they are determined to push for regulation even when the industry presents evidence that there is little risk. Outside peer review panels, she says, would be a check that "regulators with excessive zeal are not

*Word had already reached Capitol Hill about meetings Rita Lavelle, the chief of the EPA's superfund program, was having with her former employer, Aerojet-General Corporation, and other companies to work out deals that would lessen their liability for cleaning up hazardous waste dumps.

controlling the process." Schotland says that the panels would have to be carefully chosen to eliminate the potential for conflicts of interest, but the House in the 104th Congress rejected attempts to incorporate into its plan rules that would prevent manufacturers from stacking the peer review panels with their scientists.

The most important element of the Contract With America plan, to the drafters, at least, was to require rigorous cost-benefit studies before any new regulation is proposed. If manufacturers disagreed with an agency's economic calculations, they would be able to take the federal government to court and ask a judge to decide whose numbers should apply.

The implications of such a change are enormous. Federal departments and agencies have done cost-benefit analyses of their decisions since at least the late 1970s, and the industry has always challenged them. But the courts usually have been able to put aside these theoretical arguments about costs and benefits on the ground that safety comes first. As a result, judges have allowed the Occupational Safety and Health Administration to curb cotton dust on the floor of textile mills, in order to protect workers from brown lung, and to limit vinyl chloride fumes in the plastics industry, in order to save laborers from the risk of cancer—despite protests from the affected industries about the costs.

Under the Contract With America plan, judges would have to worry about the costs of all environmental, health, and safety laws. Every law would have a little bit of the Toxic Substances Control Act built into it, and, as in the asbestos decision of 1991, it would be difficult for federal regulators to convince the courts to allow regulations to be put into place.

But the public gradually turned against the plan that House Speaker Gingrich and his fellow leaders in the 104th Congress devised. They may have called their bill the Job Creation and Wage Enhancement Act, but the agreeable rhetoric ended there. Gingrich derided the EPA as a "job-killing agency'" that issues "incredibly destructive'" regulations. Majority Whip Tom DeLay, a former bug exterminator in his home state of Texas, went even further, branding the EPA a "Gestapo agency."

The cost-benefit bill was approved by the House, but a coalition of Democrats and moderate Republicans prevented the Senate from voting on the measure in the summer of 1995 by threatening a filibuster. Even then–Majority Leader Robert Dole, who was chief sponsor of the bill in the Senate, could not rally enough support to move the legislation forward. By early 1996, House Republicans had begun to abandon their leadership's hard line, with 30 members signing a letter to Gingrich protesting that the party "has taken a beating this year over missteps in environmental policy."

Supporters of the cost-benefit regulation plan blamed the harsh rhetoric of the House leadership for its demise. They also accused environmentalists and the Clinton administration of mischaracterizing their views. DeLay said that Vice President Albert Gore Jr.'s statements that the GOP was trying to repeal environmental laws was "an out-and-out lie." Republicans, DeLay said, would simply like to update the laws.

"One thing that is extremely difficult to get out of the system is the politics," says Jerry Dunn, Hoechst Celanese's vice president for environmental, health, and safety affairs. "As long as we have the attitude of us versus them, it's going to be difficult to pass any regulatory reform or to make any progress in those areas."

By May 1996, the rhetoric was toned down substantially, but leaders of the 104th Congress still had big plans for making the environmental system more industry-friendly. Gingrich unveiled a "New Environmentalism" policy, in which he announced that "the goal of Republican environmental policy is to create a cleaner, safer, healthier environment for all Americans." In a speech to the National Association of Manufacturers the next month, DeLay vowed to "push on" in the 105th Congress toward making the regulatory system more "flexible." If Dole wins the presidency and Republicans hold on to their majority in Congress, he told the manufacturers, "you're going to see the damnedest thing next year that you've ever seen in your life."

6

Justice Denied

Natalie Harrison* watched her three-year-old son double over, wheezing as usual, in the yard behind her trailer home. His coughs shattered the vision she once had for her family's life. She and her husband, Hank, had dreamed of having many children and seeing them run and play on a big plot of land like this. That's why they had chosen to buy a manufactured home near Houston back in 1989. It was 12 rooms and only $45,000—big enough and inexpensive enough to make a large family affordable.

But Natalie's son sniffled and sneezed constantly and never got better. She took him to the emergency room nearly every weekend. Her five-year-old girl had different problems. She was dizzy and sick to her stomach frequently and, to her mother's horror, had begun to lose her hair. What was going on?

For two and a half years Natalie asked doctors that question, but they had no answers. An allergist asked her if she was vacuuming the house properly and, because the boy's throat seemed so dry, whether the level of humidity could be increased. When her son developed a rash on his legs, the doctor said that she should not allow him to wear cotton underwear anymore. "It hurts me to say that, even now," Natalie says.

One day, Hank Harrison came home from work to find two vaporizers and a humidifier running, with Vicks Vapo-Rub. "Boy, it

---

*The names have been changed at the family's request.

143

smells like a hospital in here," Natalie remembers her husband saying. The memory evokes anguish, now that Natalie knows what made her son sick: "I didn't know I was adding to his problem. I thought I was opening him up."

Natalie stumbled onto the source of her son's medical problems by accident. An acquaintance, hearing about her family's problems, told Natalie she'd heard that mobile homes "have something in them."

"If you're concerned," the acquaintance asked, "why don't you move him out?" Nothing else had worked, so Natalie decided to spend a few days at her mother's house. After just three days away from home, young David was able to breathe without a respirator for the first time in nearly three years.

Natalie ultimately learned that the "something" in her mobile home—and others like it—is formaldehyde gas.

The Harrison family had overcome a major hurdle in discovering that chemicals in their home were making them sick. From the farms of the Midwest to the high-rise apartments of Manhattan, untold numbers of families suffer from similar illnesses without ever knowing why. Chemical manufacturers fight to smother the horror stories that rise like brushfires and illuminate the damage resulting from their products. The victims who do break through the silence can expect to be beaten back— first with skepticism, then, perhaps, with money, and, if all else fails, with stinging personal attacks.

The First Hurdle: Finding Out

The people of the hardscrabble towns of southern Iowa did not believe the scientist. Peter Isacson, the chief epidemiologist at the University of Iowa's College of Medicine, told them in 1989 that their babies were almost twice as likely to be born underweight than those of other Iowans. He also told them the most likely cause: high levels of atrazine in their drinking water. "I thought the women down in southern Iowa should make up their own minds," says Isacson, who is now retired and living in New Mexico. "They had a right to know. It turned out the people down there didn't want to hear it."

One reason was that the water district, one of rural America's largest, was a point of pride for the region's residents, who are among the poorest in the state. For decades, they had relied on bacteria-laden water from shallow wells—yet another indignity for a region that lacks the rich black soil the rest of Iowa enjoys. In 1971, engineers dammed the Chariton River and created 11,500-acre Lake Rathbun; six years later they built a $27 million water treatment plant and began distributing the lake water to about 13,000 people in 12 counties, including two in Missouri. Now a scientist was telling them that the water they were so proud of was hurting their babies.

Regional pride was not the only reason why the farm families did not want to hear Isacson's message. Like other victims of toxic exposures, the people of southern Iowa had been influenced by chemical manufacturers, whose first line of defense is to keep users in the dark about the dangers that their products pose.

Atrazine was the reason Isacson became interested in Lake Rathbun in the first place. Spring rains drive the weed-killer off farm fields and into the lake, where it readily accumulates in Rathbun's unusually still waters. In the late spring, atrazine levels in the lake frequently exceed the Environmental Protection Agency's safety standard of three parts per billion, although the levels usually do not stay that high long enough to violate the standard. (A reservoir is considered to be in violation of the EPA rule only if at least two water samples, collected four times per year, contain more than three parts per billion of atrazine, or if a single sample contains more than twelve parts per billion.) Another reason that Isacson was interested in Lake Rathbun is that the lake is the only source of drinking water in the 12-county region, which would make it the ideal place to study. "People are constantly talking about environmental risks, and the outcomes of being exposed to pesticides at relatively low levels," says William Hausler Jr., who worked with Isacson on the project as the director of the state hygienic laboratory at the University of Iowa. (He has since retired to Florida.) "Well, here was a tremendous geographic area for research."

As soon as Isacson reported his initial findings in 1989, his work was roundly attacked by chemical manufacturers and farm groups. "They started making disparaging remarks about our quality control and releasing commentary against my laboratory—the state laboratory," Hausler says. The companies also pointed out—correctly—that Isacson had failed to account for smoking patterns, income, levels of prenatal care, and other factors besides herbicides that might explain the disparity.

So Isacson and his collaborators did a more sophisticated analysis. This time they compared the Lake Rathbun communities with other towns in southern Iowa that had similar incomes and similar patterns of prenatal care and tobacco use. The results were virtually unchanged: The rate of underweight births was still almost twice as high in the towns around Lake Rathbun. Retests by the University of Iowa Hygienic Laboratory continued to show that atrazine levels in Lake Rathbun were high. "Every time we tested water in Rathbun we could find it [atrazine]," Keith Cherryholmes, who was Hausler's assistant at the time, says. "There were frequent violations."

But even the second study did not have much of an impact. As a July 1991 story in the *Des Moines Register* noted, Isacson's findings seemed to cause "barely a stir."[1]

One reason may have been that pesticide manufacturers had an influential ally in Kenneth Owen, the executive director of the Rathbun Regional Water Association. "I don't have a fear in the world about people drinking our water," Owen told the *Register*. Not only was Owen in charge of the water supply system, he was also a former state secretary of agriculture with long-standing ties to the state's agribusiness establishment. To this day, he regards the pesticide manufacturers as the good guys in the Rathbun controversy and Isacson (and all the other scientists who collaborated with him) as self-promoting exaggerators.

"The chemical companies, they were there to solve the problem," Owen says. "They were as cooperative as they could be. I've never seen them try to avoid anything or try to hide anything. We feel as safe as can be about our water. I don't think anybody around here has ever worried about it one iota."

At meetings in Des Moines and in Centerville, the biggest town in the Rathbun region, Owen repeatedly challenged Isacson, painting him as an outsider and a threat to the region's struggling economy. Representatives of Ciba-Geigy, Monsanto, and the National Agricultural Chemicals Association did the same. "I can remember one evening meeting in the state capitol," Hausler says. "The Monsantos and Ciba-Geigys were there, and they were really playing to the farmer, and to the better life that those who are growing crops now have because of chemicals. It was clear that they had the public with them. We were butting our heads against a stone wall."

By the time the study was presented at the Society for Epidemiologic Research's 25th annual meeting in Minneapolis in 1992, Isacson's findings had been strengthened further by another University of Iowa researcher, Ron Munger, who is now at Utah State University. Using hospital records, Munger compared birth-defect rates in 18 communities around Lake Rathbun to those elsewhere in Iowa from 1983 to 1989. The results: The Lake Rathbun babies were 6.9 times as likely to have short limbs, 3.5 times as likely to have genital defects, 3.1 times as likely to have heart defects, and 2.6 times as likely overall to have any type of birth defect. Water samplings taken by Hausler's laboratory, meanwhile, showed that drinking water supplied by the Rathbun Regional Water Association averaged 2.2 parts of atrazine per billion in 1986 and 1987, compared to an average of six-tenths of a part per billion in surface waters used for drinking in the rest of the state.[2]

The Rathbun studies cried out for a follow-up, but there has been none. The EPA, the Centers for Disease Control and Prevention, the National Cancer Institute, and the Public Health Service's Agency for Toxic Substances and Disease Registry have all expressed interest at one time or another but ultimately bowed out. All four key researchers— Isacson, Munger, Hausler and Cherryholmes—have left Iowa. And the researcher in the best position to conduct a follow-up study says that it is a low priority because of the high cost and lack of interest in local communities.

"We as scientists may be partly to blame for that, for failing to follow up," Dr. Charles Lynch, who now heads the University of Iowa's cancer epidemiology research program, says. Testing a water sample for pesticides costs more than $100, and a more sophisticated Rathbun study would involve thousands of such tests as well as in-depth questioning of hundreds of residents. If local residents were pushing for answers, Lynch says, a public health agency might be willing to foot the bill. "There has not been a lot of receptivity among the population in Rathbun," he says. "In general, there wasn't a great impetus that we were hearing to do more."

Ciba-Geigy scientists agree that the Rathbun studies should have been followed up but say they are confident atrazine does not cause birth defects. "The compound has been used for a very long time, and there haven't been claims along those lines," says Darrell Sumner, Ciba-Geigy's manager of health and safety issues. "This is the only accusation. Atrazine has been tested through a whole series of animal reproductive studies. Atrazine does not produce any reproductive toxicity. It does not produce birth defects."

The people of Lake Rathbun may not have wanted to hear about the herbicide threat, but they were at least able to choose for themselves to ignore it. Many others suffer from the ill effects of chemicals without ever even knowing that they have been at risk. How many others? We will never know, of course. But there is ample evidence that perchloroethylene, formaldehyde, alachlor, and atrazine are an everyday presence in our lives—and that those four chemicals represent just a tiny fraction of all the toxic chemicals Americans are exposed to every day.

In California, where an unusually strict state law has required that most groundwater systems test for the presence of chemicals, more than 35 percent of the 750 wells in the huge Central Valley are known to contain perchloroethylene, many with levels above the state drinking-water standard of no more than five parts per billion. State officials have been able to trace the pollution to the source at 21 of the wells; 20 of the sources were dry-cleaning shops. Furthermore, the flow of perc did not follow the slope of the land but the trail of the sewer line, meaning that the perc had been dumped down the drain and seeped into the soil through cracks and fissures in the sewer system.[3]

State and city officials have made similar discoveries in Florida, Maryland, Massachusetts, New York, and Ohio. "Disposal practices of the industry may well mean that a perc plume exists underneath virtually every dry-cleaning facility in the country," Bonnie Rice of Greenpeace says.[4]

Perc is in the air, too. A battery of studies that the EPA conducted from 1987 to 1993 in California, New Jersey, and North Carolina showed that perc was among the toxic pollutants found at the highest concentrations in urban air.[5] In New York, the state that has studied the problem the most, an estimated 170,000 residents are exposed to high perc concentrations in the air because they live or work near dry-cleaning establishments.[6] Ever buy butter or other fatty foods from a convenience store that is next door to a dry cleaner? One EPA study showed high levels of perc in cheese, butter and margarine, ice cream, meats, and other high-fat products; the perc fumes dissolve and accumulate in such products as efficiently as they dissolve grease from clothing.[7]

Formaldehyde is everywhere, as the industry itself proclaims with pride. It is an invisible presence in every home or office—a gas that rises from cabinets, tables, wall paneling, wood molding. There is no way to get rid of formaldehyde gas except to let it dissipate over time, until a room is remodeled or refurnished, beginning the cycle again. While many people do not sense the gas at all, others cannot breathe when it is around. It can make those people asthmatic, disoriented, and nauseated; for some people who are very sensitive to the chemical, it can produce crippling, asthmatic effects for the rest of their lives. The EPA estimates that 10 to 20 percent of the U.S. population, including asthmatics, may have hyperreactive airways that may make them especially susceptible to getting sick from formaldehyde.[8] And always lurking down the road—even for those who are not immediately sickened by it—is the possibility that formaldehyde will cause cells in the lung, nose, throat, or elsewhere to mutate, multiply, and inflict the body with cancer.

The residents of southern Iowa are not the only people at risk from herbicides. The biggest single study, conducted in the Midwest by the U.S. Geological Survey in 1993, found weed-killers in 95 percent of the 77 reservoirs tested and 97 percent of the 161 sampling stations along rivers and streams.[9] Based on that study, as well as on other data collected by the EPA and state governments, the Environmental Working Group concluded that about 11.7 million people in the corn belt states and Louisiana drink water drawn from sources that contain atrazine, alachlor, or three other major herbicides.[10] Weed-killers are also found at lower levels in all five Great Lakes and the Chesapeake Bay watershed, which are sources of drinking water for an additional 12 million people. Farmers, who tend to have the greatest exposures, are particularly vulnerable because they often do not wear gloves when pouring weed-killers, use their bare hands to handle spraying equipment, and sometimes even blow out clogged nozzles with their mouths.[11]

Such widespread exposure to so many known or suspected carcinogens is inevitably taking a toll on human health. In September 1994, a National Cancer Advisory Board impaneled by Congress reported that cancer will strike a third of the nation's population and kill one in every five Americans, surpassing heart disease as the predominant cause of death in the United States in the coming years. One of the panel's chief conclusions was that current laws, public policy, and government regulation undermine cancer prevention, treatment, and control efforts.

"Lack of appreciation of the potential hazards of environmental and food source contaminants worsen the cancer problem and drive up health care costs," it said.[12]

Even taking into account the increases that could be due to improved diagnostic techniques and the fact that people are living longer, it is clear that some cancer rates have risen mysteriously—including those for non-Hodgkin's lymphoma, for brain, kidney, and testicular cancer, and for certain types of breast cancer. All have been linked in animal or human studies to pesticides, solvents, and other chemical products. Yet the manifold health effects of smoking, genetics, diet, and a host of other factors ensure that there is almost no chance that any particular case of cancer will ever be traced back to its cause.

Still, almost no chance is too much of a chance for chemical manufacturers. So they have worked to make the odds even longer whenever one victim happens upon success in tracing the chemical trail.

Only a few lawyers, victims, and manufacturers, for example, know the full story about carbonless copy paper. According to Thad Godish, the director of Ball State University's Indoor Air Research Laboratory, who has testified on behalf of such victims, hundreds of lawsuits across the country have been filed by people who got sick or developed multiple chemical sensitivity from being exposed to this formaldehyde-impregnated paper; all of them, he says, were settled with confidentiality agreements.

The available court records show that at least nine such cases were in the federal courts in 1994, when a special appeals panel blocked a joint effort by the plaintiffs to bring their cases together as one to make the litigation less costly.[13] In early 1996, although most of the nine cases still were open, their lawyers believed that because Appleton Papers, Inc., the nation's biggest manufacturer of carbonless paper, had obtained a gag order, there was little that they could say publicly about the cases. Under the rules used in most courts, judges can refuse to agree to such an order unless the parties show good cause—for example, that trade secrets might be inadvertently revealed. However, judges routinely approve such orders if the defendant (which wants secrecy) and the plaintiff (who wants money to deal with immediate medical problems) both agree to close the record as a condition of settling the case.

Evan T. Lawson, a partner in the Boston firm of Lawson & Weitzen, agreed to speak in general terms about a carbonless-copy-paper case in which he obtained a settlement. His client was a woman who believes that she had an allergic reaction and developed multiple chemical sensitivity at her job, where she was responsible for separating and filing carbonless forms all day long. Carbonless paper has a backing in which an ink is contained in microcapsules that break under pressure. Formaldehyde was used as a stabilizer in the ink in the 1980s, Lawson says. Not only did the paper give off formaldehyde gas, but when carbonless forms were separated, they created formaldehyde-laden dust. By 1996, he says, Appleton Papers had either reduced or eliminated formaldehyde in its products.

But many people, including his client, have lingering illnesses that they attribute to their past exposure to widely used carbonless products. "It makes it hard for them to function in an urban, interior environment," Lawson says. Although he had only one carbonless-paper case open in 1996, he said that he has been approached by other potential clients but rejected their cases. That was not because they lacked merit, he says, but because he knows how difficult such cases are to win.

"There's a whole little dance we do," Lawson says. "The industry says it could have been this, this, this, or this. It could have been the cooking oil that burned in your kitchen ten years ago. Well, yes, it could have been—but what is it likely to have been?" (Lawson says that he will not represent a smoker with a complaint of formaldehyde injuries, for example, because tobacco smoke contains formaldehyde.)

Richard Lippes, a lawyer in Buffalo, New York, who worked on the notorious Love Canal pollution case, won settlements in hundreds of cases against manufacturers of the formaldehyde that went into foam insulation—Borden, DuPont, Georgia-Pacific, and others—in the 1980s. Almost all the cases were quietly settled out of court; most plaintiffs, in fact, agreed to settle before the filing of a lawsuit so that they could pay their immediate medical bills. They typically sign agreements in which the chemical manufacturers admit no harm and the plaintiffs promise not to disclose the terms of the settlement or the documents that they have gathered to support their cases.

Lippes and others admit that while signing such an agreement is a difficult decision, they are sworn to represent their clients, not the public at large. The result is that consumers remain uninformed, and the next victim who decides to fight manufacturers on the same issue will be forced to begin the fight anew. Godish says that he has testified on behalf of plaintiffs in nearly 300 formaldehyde cases; only 16 of those, he says, have gone to trial.

The confidentiality agreements take a toll even on those who sign them. "It's victimizing the people all over again—it's invalidating what happened to them," says Connie Smrecek of Minnesota, a mother of two who became severely ill from particleboard that was used in the remodeling of her home and went on to become one of the most vocal anti-formaldehyde activists in the United States.

In her many conversations with fellow formaldehyde victims, Smrecek has found that signing such agreements are among the most painful decisions they faced. She recalls a woman who, after being sickened by exposure to formaldehyde in her office, was asked by her employer to sign a confidentiality agreement to get worker's compensation. "She was concerned that when she goes to the physician, if she by chance lets something slip, does this mean her workmen's comp will be taken away?" Smrecek says.

Secrecy colors almost everything manufacturers do when there is a chance that they might be held accountable for the dangers posed by their own products, as one of the most famous formaldehyde cases shows. Manufacturers kept the formaldehyde threat hidden from the Pinkerton family of Centralia, Missouri, and that secrecy later emerged as a key reason why the Pinkertons won a $17 million jury verdict.

"It's a very sad situation," Walters says. "One of the things that was in this case: The purveyors of formaldehyde are not quick to acknowledge that it is a sensitizer."

Walters is speaking about one of the most devastating impacts of chemicals like formaldehyde: People who are sensitive to its effects, and are exposed to enough of it to make them sick, sometimes develop similar allergic reactions to *any* chemical in their homes or offices. And because chemicals are everywhere in modern society, people with this multiple chemical sensitivity find it very difficult to function at all. During the Pinkerton trial, Walters showed that Georgia-Pacific and Temple Industries had developed warnings for their factory workers that showed they knew formaldehyde had this potential.

"But they never published much of it," Walters says. "Even the warnings required by the government to be on particleboard do not mention that formaldehyde is a potential carcinogen or is, in fact, a sensitizer." In keeping key information from the public, Walters adds, "they've done a nice job."

After a large verdict, plaintiffs usually can expect months and sometimes years of negotiation and wrangling among the attorneys as the defendant chemical company attempts to have the verdict overturned. In the Pinkerton case, Georgia-Pacific and Temple Industries elected not to take the matter further; a new trial surely would have paraded before the public the potential fatal effects of multiple chemical sensitivity and the death of Mary Pinkerton. The manufacturers settled with the Pinkertons for an amount that can be presumed to be somewhat less than the $17 million that the jury said they should have. The exact figure, of course, remains secret, as does any further information about the case.

It does not happen often, but sometimes the veil of corporate secrecy is pierced, stripping away the barriers that have shielded manufacturers from having to pay for their products' harms.

That is what happened when Houston lawyer Robert Bennett walked into the offices of the Texas Manufactured Housing Association. Bennett had begun handling cases on behalf of formaldehyde victims in the early 1980s, after the *Houston Post* ran a series of stories on the health problems of people who lived in mobile homes, which typically have the worst formaldehyde problems. Most of his cases had been settled quietly out of court.

By 1991, Bennett had been sparring for years with the mobile-home industry over what internal documents he had a right to see. Finally, Will Ehrle, the general counsel and president of the Texas Manufactured Housing Association, told Bennett to come over to his office at 10 o'clock the next morning so that they could go through the documents and try to break the impasse.

When Bennett arrived, Ehrle was not in the office. Ehrle's secretary, however, "demonstrating an unusual degree of hospitality," Bennett says, invited him into the room where the documents were filed and left him alone to read whatever he wanted. Later, when the secretary was leaving for the day, she told Bennett that he could stay as long as he turned out the lights when he left.

"I was like a busman on a busman's holiday," recalls Bennett, who spent six hours at the photocopying machine. He knew that the nearly dozen years' worth of minutes of the Formaldehyde Institute and the letters and memos passed among the trade associations, Borden, Celanese, DuPont, and Georgia-Pacific were as valuable as gold.

In 1992, the U.S. Supreme Court set a difficult hurdle for victims to win punitive damages from manufacturers, in a case involving Rose Cipollone, a New Jersey woman who sued tobacco companies over her lung cancer.[14] Punitive damages are the awards—sometimes in the tens of millions of dollars—intended to punish defendants for their wrongdoing and to deter them, and others, from repeating their injurious activities. In the Cipollone case, the Supreme Court held that plaintiffs seeking punitive damages from manufacturers must prove that the manufacturers deliberately withheld information or deceived the public.

That is why there has been such a fuss over the internal records of tobacco companies. It's common knowledge that cigarettes cause cancer, but the courts don't know it, in the peculiar manner of blind justice that has evolved in the United States since the Cipollone case. For a plaintiff to obtain internal records—such as the ones that Bennett now gathered on the formaldehyde industry's lobbying and litigation program—is rare indeed. Although the law, in theory, requires manufacturers to turn over all relevant records, they do not have to give over documents that are covered by "attorney-client privilege." Bennett could see in the records he read as he passed them through the copying machine that the formaldehyde manufacturers had taken great care to ensure that lawyers had signed off on all of their documents, including scientific work and draft press statements.

Now that a lawyer had permitted him to see the documents, Bennett also knew, the attorney-client blanket of secrecy would be lost. The next day he filed the records in court to ensure that they would remain public. Ehrle's organization fought to seal the records, but after a hearing at which both Bennett and the hospitable secretary testified, the judge ruled that the documents had been provided voluntarily. No privilege applied.

Bennett credits the documents for the $600,000 settlement he quickly obtained on behalf of the George Slaughter family for their nine-year-old son, Jason, who developed reactive airway disease from formaldehyde exposure in their mobile home. He then took a group of about a dozen cases into mediation—a "private" sort of courtroom where the proceedings are not on the record—and obtained settlements that averaged about $25,000 apiece. He prepared a videotape to show the mediator that summed up his argument with a word of self-congratulation. Bennett's firm had done more than "any law firm in the country to establish [that] a conspiracy existed to keep the public from knowing the true dangers of formaldehyde."

The documents also allowed Bennett to reach beyond the mobile-home dealers into the "deeper pockets" of the corporations that had set up and controlled the Formaldehyde Institute. It was an ironic twist that the mistake of the Texas Manufactured Housing Association—allowing a mountain of internal documents to fall into the hands of a plaintiff's lawyer—had opened the trail of blame to the Formaldehyde Institute; for years, albeit behind closed doors, the mobile-home dealers had grumbled that they were bearing the brunt of liability for a product—formaldehyde—that they did not make.

The consequences for the Formaldehyde Institute were fatal. Because its liability insurance company had to pay out so much in claims, it did the same thing that an auto insurance company does to drivers who have shown themselves to be risky customers. The insurance company jacked up the Formaldehyde Institute's premiums.

"The only rates we could get were so high we couldn't afford to keep the institute together," says Jerry Dunn, Hoechst Celanese's vice president of environmental, health, and safety affairs. Also, he said, the institute reevaluated its agenda. It had just completed working with the Occupational Safety and Health Administration on its development of a formaldehyde standard. This was the intensely negotiated regulation that included carefully worded warnings for workers: that formaldehyde was an animal carcinogen at high levels, an irritant at lower levels, and a possible irritant at still lower levels. It was a success for the industry, because workers would not be warned at all that formaldehyde was a human carcinogen. "Our driving force had somewhat decreased," Dunn says. "But the major reason [for shutting down the Formaldehyde Institute] was monetary and insurance."

Soon, in Washington, the EPA regulators who were fighting a difficult, if not losing, battle every day over formaldehyde were stunned to hear, as one EPA staffer put it, that "some Texas lawyer put the Formaldehyde Institute out of business." The Formaldehyde Institute filed papers of dissolution on July 30, 1993, nearly a dozen years after it was born.[15]

The Second Hurdle: Getting to Court

There is practically nothing more important than basket weaving to the culture of the Yurok Indians of northwestern California.

"We are a tribe that centers around our baskets," says Susan Burdick, a Yurok who teaches basketry at a local college. "Our baskets are used in our sacred dances that we still do. They're used for gathering. We make caps for the young girls to dance in. We make our baskets for the world-renewal dance. Basketry is at the center of who we are as a people."

But it is also at the center of a health mystery that is both frightening and frustrating the small tribe. Burdick and others are convinced that an unusually high number of the Yuroks get cancer and respiratory diseases and suffer miscarriages, and that the reason is the heavy use of herbicides—including atrazine, 2,4-D, and Garlon-4—in the private and public timber forests in and around their reservation, where the Yuroks regularly forage for spruce roots and hazel, sugar pine, and willow sticks for basket weaving. Their exposure is alarmingly direct because they chew the sticks to soften them for weaving. "Our mouth," Burdick says, "is like our third hand."

But there is no way for Burdick and the rest of the Yuroks to prove, scientifically or legally, that herbicides are injuring them. The problem is that only a few hundred of the 3,000 Yuroks actually live on the tribe's vast reservation. That is not nearly a dense enough population of Yuroks for statisticians to prove that an unexpectedly high number of them are dying prematurely.[16]

"There does seem to be a high incidence of cancer and respiratory diseases on the reservation," Sara Greensfelder, the executive director of the California Indian Basketweaver Association, says. "But because it's a rural area with a widely scattered population, epidemiological studies wouldn't be valid. So all we have are very sketchy and informal health surveys."

The limitations of epidemiology are just one of the problems that keep victims out of court, even when there is evidence that chemical exposures may be the cause of their injuries. Legal standards of causation are so tough to meet in chemical exposure cases that often the biggest problem victims face is finding a lawyer to take their case. Because plaintiffs can rarely afford to pay by the hour—toxic-injury cases often cost millions of dollars to litigate—they have to persuade lawyers to take their cases on a contingency basis. In a contingency case, a lawyer might end up with 30 to 40 percent of a verdict or settlement, or with nothing at all if the case is lost.

Raymond Booth III, a Jacksonville-based lawyer who specializes in asbestos litigation, decided that the potential rewards did not justify the risks of taking on the case of an alachlor-using farm family from nearby Lake City whose child was born with severe birth defects.

"The tough part of these cases is proving the causation," Booth says. "Everybody's familiar with asbestos litigation, but there's some unique things about asbestos. You get a unique disease, asbestosis, which gives you some unique X-ray patterns. You get mesothelioma, which virtually everyone will agree is caused by asbestos. You also do get lung cancer and other cancers of the digestive system, but generally those folks have died by the time you get to trial and the pathologists can take out the malignant tissue and try to pinpoint the site of origin, look for asbestos fibers there, and use that to make a causal connection. But when you're dealing with birth defects, you don't have those other means. It's very, very difficult to show causation."

Booth decided not to take the case after a doctor said that the child's problems—cerebral palsy with spastic paralysis—could have been caused by something other than the heavy use of alachlor on the family farm.

The legal system places such a heavy weight on causation that, in the world of chemical illness, the presence of fingerprints has become far more important than the severity of the crime.

When thousands of people developed respiratory illnesses after they installed formaldehyde foam insulation in their homes in the late 1970s, they received immediate attention in the news media and from federal regulators, who tried to ban the product. Mobile homes have never received similar attention, even though they always had far higher levels of formaldehyde gas than did other homes with formaldehyde foam insulation.

"That's because people were literally chased out of their homes," Godish says. "It made for a good story, because one day their home was healthy, and the next day it wasn't." Lippes, the Buffalo lawyer, says that he received as many as 80 telephone calls a week after ABC-TV's *20/20* broadcast a segment on the problem.

But for people who were actually getting higher exposures from what seemed to be a common, benign product also mentioned on the *20/20* show—particleboard—the burden of proof was far more intimidating to victims and lawyers alike.

Michael Hart of Turnipseed & Associates, a law firm in Columbia, South Carolina, recalls the story of a 31-year-old woman who became sick after just one night in her new mobile home in 1991. Her stomach and head ached, and her face was swollen and red. By the time the illness was traced to formaldehyde, her lungs had filled with fluid, her blood was dangerously thin, and her kidneys had shut down. The woman's doctor, however, told Hart that one possible explanation for her kidney failure was that she might have taken too many pain relievers. Without a doctor to make a definitive statement on the issue of causation, Hart said, his law firm could not take her case.

James H. Davis, a Los Angeles lawyer, usually rejects toxic-tort cases because they are so difficult to prove. But in the mid-1980s he agreed to represent an immigrant laborer who had suffered a stroke and was left paralyzed and unable to speak. The man's daily job had been to operate a mechanical spraying machine for a California railway company; as the nozzles rotated, atrazine would regularly splash onto his pants legs. But a jury refused to hold the railway liable because, it concluded, the man's paralysis may have been caused by a viral infection.

"It's a real problem when you're talking about somebody developing a latent disease, cancer, or something like that," Patti Goldman, a lawyer with the Sierra Club Legal Defense Fund, says. "The causation seems to be clearer when it's crop damage, or the acute dermal problems or eyesight problems that farmworkers might have. But farmworkers aren't likely to bring cases, and it's very difficult to establish causation for the more long-term health problems."

The sophisticated epidemiologic studies that might prove such links are expensive—and industry, for obvious reasons, is not eager to bankroll them. It was not until 1994 that the first definitive human studies on the effects of perchloroethylene were published in the medical literature.

Even when a victim manages to persuade a lawyer to take on a chemical-exposure case, there is no guarantee that a judge will allow it to be heard. Toxic-tort lawsuits are generally thought to be much stronger if they are combined into one as a "class action," because the manufacturer is denied its strongest defense: tearing apart a single plaintiff's claims. But as Lippes was preparing a class-action lawsuit on behalf of 70,000 owners of homes with formaldehyde foam insulation, he did not know that the formaldehyde industry's top strategists—officials of Borden, Celanese, DuPont, Georgia-Pacific, and other manufacturers—were meeting in the New York City offices of one of their law firms.[17] Sara Schotland told the group that a lawsuit filed on behalf of a small group of victims in Pennsylvania was "being moved ahead on as fast a track as possible with the hope that a favorable decision in that case could be used to dismiss the other three pending court actions."[18]

Why did Schotland smell victory in Pennsylvania? It could be because the judge in the case had already rendered some extremely favorable rulings. It could be because Schotland had effectively isolated the Pennsylvania victim's attorney from working with other lawyers by negotiating a deal that he could not show the documents he obtained during the pretrial exchange of evidence to any other lawyers for victims. What is important is that the formaldehyde manufacturers had the resources to mount a nationwide war plan that not even a "class" of victims could hope to match.

Two years later, citing a Pennsylvania decision in favor of the formaldehyde industry (but apparently not the one that Schotland had talked about), a New York judge refused to allow Lippes's class of formaldehyde foam victims to go to court en masse.

The Final Hurdle: Getting Justice

Members of the Graham family of tiny Norton, Vermont, thought that they had a strong case when they filed suit against the Canadian National Railway in 1986. They had already cleared the two hurdles that trip up most other victims. They had traced their injuries back to a toxic exposure, and they had found a lawyer to take their case to court.

For more than a decade, companies hired by the railway had repeatedly drenched with herbicides a train track and adjacent right-of-way that cuts through the middle of the Graham family's 13-acre farm. And for more than a decade, life on the little farm had been one horror after another.

Every time the motorized spraying machines would move down the tracks and across the farm, brown and white powder would drift from the right-of-way into the barnyard, the fields, the pond, and even into

Brigitte Graham's garden. On hot summer days, the pond would bubble and foam. On rainy days, a white foamy substance would fall from the trees. Dead birds and bees littered the area near the tracks; many of the surviving birds had skin and eye problems and were missing feathers. Trees on the farm lost their leaves, as did vegetables in the garden. One winter, the family cow's fur fell out, and then she got abscesses in her feet and blood in her milk. Brigitte Graham, who would spend hours working in her vegetable garden, developed raw open sores on her hands. All five family members would get rashes and have trouble breathing in the days following a spraying. The respiratory problems did not go away for Brigitte's husband, Edward, who also developed skin cancer. Repeated bouts of pneumonia left him bedridden—and, in January 1990, finally killed him.

By the time the case was ready for trial shortly after Edward's death, the Grahams had put together what looked to be a formidable case. They had had their well water tested by the lab director of the toxicology department at Cornell University's College of Veterinary Medicine, who found traces of three herbicides used by the railway: atrazine, diuron, and bromacil. The family's veterinarian, an eyewitness to many of the problems on the farm, had taken blood samples from livestock on the farm and found traces of herbicides, too. And the family members themselves had been examined by Benjamin Hoffman of Exeter, New Hampshire, one of New England's leading environmental toxicologists, who concluded that their skin rashes were consistent with known symptoms of overexposure to atrazine. Hoffman, like the Grahams' own doctor, believed that the family's medical symptoms were probably caused by chemical exposures, not by animal waste bacteria, cigarette smoking, poor hygiene, or acid rain—all alternative explanations that had been suggested by the railroad.

But in April 1990, after a five-week trial, the case was dismissed. The reasons U.S. District Judge James Holden (now retired) listed for throwing out the case starkly illustrate the crushing legal burdens that victims of chemical exposure are so rarely able to overcome.

The railway, Holden concluded, was indeed negligent. The spraying firm it had hired not only had violated the law by failing to get a license from the Vermont Agriculture Department, it also had ignored safety warnings printed on product labels and failed to alert the Grahams and other nearby landowners that hazardous chemicals were being sprayed. "It is clear," the judge wrote in his order, "that the undertaking was dangerous."

The problem for the Grahams, the judge wrote, is that there was insufficient evidence that the company's negligence had caused the family's injuries. "Proximate cause," he wrote, "is the linchpin that fastens wrongful conduct to legal liability." The Grahams' symptoms—their rashes, their headaches, their breathing problems, and even their cancers—could have been triggered by any number of causes. The Cornell test had found weed-killers in the family's water, but the levels were within EPA guidelines, and many other tests conducted by state agencies had failed to detect any herbicides in the water. There was no "reasonable medical certainty," the judge concluded, that the chemicals sprayed by the railway had caused the family's problems.[19]

"The whole thing was just so frustrating," Brigitte Graham says. "There were so many chemicals sprayed on the tracks for so many years, and we had to try to prove which chemical and what amount did what damage to us."

An appeal was out of the question. The Grahams had spent $150,000 pursuing the case and were already in debt, having exhausted their life savings as well as a small settlement they had received in a separate lawsuit against one of the spraying companies hired by the railroad. "We couldn't keep going," Elizabeth Graham, Brigitte's daughter, says. "We just decided we had enough."

Today, things are different at the Graham farm. Most of the animals have been sold, and Brigitte Graham no longer works in the garden. After the trial, the railway stopped spraying along the tracks that bisect the farm. Ironically, Elizabeth Graham says, there are not any more weeds on the right-of-way now than there were during the years of spraying.

But the horrors continue for the family. In 1994, Brigitte Graham was diagnosed with an inoperable brain tumor. She is plagued by headaches and blurry vision, and her long-term prognosis is uncertain. "I don't know what's going to happen," she says. "It just seems like there's nothing we can do about it. We tried, and we lost."

The chemical industry's campaign to beat back toxic-tort litigation has a crucial ally in the so-called preemption clauses of the nation's major environmental laws. FIFRA, for example, specifies that only the EPA, not state and local governments, can require special warnings or packaging for pesticides. Manufacturers have interpreted that clause so broadly that they now routinely argue that victims cannot sue them under traditional common-law theories such as failure to warn, negligence, design defect, or breach of warranty. Every time the EPA approves a label for a pesticide, the industry argues, it has, in effect, assumed the liability for any damage that a product might someday cause.

Startlingly, judges more often than not have accepted that broad industry interpretation. Consider the case of Minas Papas of Clay County, Florida, who developed a blood disorder—aplastic anemia/pancytopenia—and became totally disabled after working for four years at a local animal shelter, where his duties included giving dogs flea baths with two kinds of insecticidal shampoo. In 1988, he sued the manufacturers of those shampoos, Upjohn Company and Zoecon Corporation, in federal court in Jacksonville. Papas's lawyers argued that the companies were liable because the labels on their products did not warn that they each contained the inert ingredient benzene, a highly toxic chemical that has been shown to cause aplastic anemia/pancytopenia. But Papas's case never reached a jury. The trial judge dismissed it on the basis that the EPA's approval of the warning labels absolved the manufacturers of liability, and appellate courts upheld the decision.

An even more flagrant example involves the death of nine-year-old Marlo Strum of Columbus, Georgia. A week before Halloween in 1987, he found some pellets in an old mayonnaise jar behind the counter at a local Boys' Club and ate them like candy. The pellets were rat poison, a blood thinner marketed under the name Talon-D. Four days later the boy's nose began to bleed; a few days later he was hospitalized and received blood transfusions. By the time poisoning was diagnosed, the antidote was ineffective and Strum died. His family's lawyers sued the manufacturer, ICI Americas, Inc. They argued that ICI knew children had been poisoned by the product in the past and could have made Talon-D safer by making its label warning stronger, by making it taste bitter, and by including ingredients that would induce vomiting if swallowed. A jury agreed and awarded his family $1,510,000. But the appellate court threw out the verdict on the basis that it was the EPA's job to determine whether a label is adequate and that, once the EPA approved Talon-D's label, ICI could not be sued for failure to warn.

The preemption factor has become much more important since the Supreme Court's 1992 ruling in the Cipollone case—the same ruling that has made it so difficult for plaintiffs to win punitive damages from manufacturers. The court's majority concluded that Congress intended for federally mandated warnings on cigarette packages to preempt state damage suits. Environmental organizations argue that Congress never had any such intention for FIFRA pesticide warnings, but judges have nonetheless repeatedly cited the cigarette ruling in dismissing lawsuits against pesticide manufacturers.

"You see the preemption argument now in all kinds of product liability cases," says Bob Shields, a prominent toxic-tort lawyer in Atlanta who represents Marlo Strum's parents and is continuing to pursue the case on appeal. "After the Cipollone case, the decisions really started piling up against you. It makes you more reluctant to go after the manufacturers. It changes, in the pesticide area, what kind of cases you will look at."

Goldman of the Sierra Club Legal Defense Fund agrees. "The biggest problems lately have been preemption," she says. "The industry's really been able to capitalize on the cigarette preemption cases. They argue that once they get a label approved, there is no duty to do a greater warning. And the courts have generally agreed with them." In fact, of the more than two dozen federal and state pesticide cases dealing with preemption that Goldman has tracked in recent years, about two-thirds have ended with rulings that FIFRA preempts state and local law.

Bennett and the few lawyers still handling formaldehyde cases are having similar problems. In 1984, the formaldehyde industry persuaded the Housing and Urban Development Department to require sellers of manufactured homes to provide buyers with a formaldehyde warning statement.[20] It may seem odd that the manufacturers would want such a stigma attached to their products; indeed, the manufactured-housing industry tried unsuccessfully to persuade the Bush administration in the early 1990s to repeal the regulation with a "Protest the Warning" letter-writing campaign to then–HUD Secretary Jack Kemp.

But for the manufacturers of formaldehyde, the warning has been a blessing in disguise because it has preempted lawsuits just as the federal pesticide law has. "It has a twofold purpose," Schotland says. "It's protective of the consumer and, secondly, it avoids excessive litigation over the warning."

Bennett, however, has several cases still open. He argues that mobile-home buyers were never properly warned of the formaldehyde problem, particularly of its potential for causing cancer. (Schotland says that the industry is "vigorously, violently" opposed to cancer labeling because it does not believe formaldehyde is a carcinogen at the levels it reaches in homes and workplaces. Bennett and his scientific witnesses disagree.)

But another important technicality now affects most of Bennett's formaldehyde work. The manufacturers have begun to argue—with some success—that he cannot bring his allegations of conspiracy into a courtroom in Texas unless he can prove that the conspiracy actually occurred in the state. Plenty of formaldehyde meetings occurred in Austin, Dallas, and Houston, and there were many letters exchanged among the manufacturers and their Texas mobile-home customers. However, he cannot prove that the national trade associations in which the manufacturers worked together were "doing business in Texas," as the Texas courts require before they take jurisdiction of a case. Bennett and his clients will have to settle for suing the mobile-home manufacturers and their trade associations and, when they can be traced, manufacturers of component parts, such as Georgia-Pacific. The rich evidence of the cross-industry meetings and talks on formaldehyde will not be a part of these cases.

Results like this are not an accident, of course. They are the product of intensive efforts by manufacturers to foment the scientific doubts that shield corporations from liability. Their defense is "very vigorous," says Eugene Brooks, an Atlanta lawyer who at one point had a regular caseload of three to four formaldehyde cases. "They play on the difficulty of the proof. 'Were the air tests properly done to prove exposure? Do you know the amount of formaldehyde in the product?' You had to find a credible doctor who would diagnose these folks. The value of a case totally depended on a doctor's testimony."

Peter Breysse of the University of Washington, an engineer who has testified on behalf of many formaldehyde victims, says: "I've listened to general practitioners get turned inside out by lawyers for the large corporations: 'You're an expert, are you? You mean you've exposed animals? You've researched people? Do you know of any other chemicals that can cause the same effects?' And usually there are. They pull out a big reference book about environmental diseases and ask: 'Do you know this author? Do you know this disease?'"

The industry's aggressive tactics extend to threatening scientists with lawsuits. Karim Ahmed, a biochemist with the Natural Resources Defense Council, had been a particularly damaging witness against Monsanto during the mid-1980s, when Canada, Massachusetts, and Wisconsin all held hearings to lay the groundwork for either banning or strictly controlling alachlor. In 1987, Monsanto sent a letter to Ahmed threatening to sue him for breaking a confidentiality agreement. The company said that Ahmed, who was then a member of the EPA's Science Advisory Board and an associate professor of toxicology and environmental sciences at the State University of New York (Purchase), was disclosing alachlor's inert ingredients, a trade secret.[21]

"It's very hard to tell whether the intimidation came as company policy or as a result of some very zealous lawyers," said Ahmed, now a consultant. "Anyway, when you work for NRDC, you get used to strong-arm tactics from industry."

The roughest courtroom tactics are usually reserved for the victims themselves. People who sue chemical companies must expect that their own lives will come under rigorous scrutiny.

In a magazine for corporate attorneys, Joseph P. Thomas, a Cincinnati lawyer, explains the strategy. He wrote, "Formaldehyde litigation presents a situation where every stone should be overturned" in looking into a victim's background. Because formaldehyde is so common, he suggests looking for other possible sources of exposure. Thomas suggests the following catalogue of areas to investigate: "A survey of the living and work environments for the plaintiff's entire life; an evaluation of plaintiff's exposure to cigarette smoke; an evaluation of personal hobbies, such as cooking, gardening, and photography; and much more.... If a thorough history is established, alternative sources will be found." Also, Thomas points out, "there may be a history of anxiety, depression, or psychosomatic disorders that can account for plaintiff's complaints."[22]

Public-relations representatives for the formaldehyde manufacturers have sometimes cringed at the hardball tactics of the lawyers. "There has been a tendency within some industries to refer to some of the complainers as unsophisticated and not really [having] a basis in fact for their complaints," according to the minutes of one meeting of the Formaldehyde Institute. "It was pointed out that not all people are that type. There are definitely some individuals who have real problems."[23]

Nevertheless, Ball State University's Thad Godish says that he has seen many formaldehyde victims put on trial, with questions about the most personal and intimate details of their lives all brought on stage. If a person smokes, he says, the settlement is likely to be significantly lower. And people who live in mobile homes may have the steepest road to justice. "People who live in mobile homes are at the low end of the economic ladder," Godish adds. "They are individuals who, compared to the general population, are the least likely to take action when they're being wronged. And they usually don't even know they're being victimized."

There have been no major lawsuits yet on dry-cleaning-related perchloroethylene injuries. But in New York City, where hundreds of dry cleaners are located in buildings with apartments or condominiums upstairs, residents have already gone public about the problems caused by perchloroethylene in their buildings only to find themselves the targets of personal attacks from chemical interests.

When the EPA held hearings in New York City on dry-cleaning risks in November 1993, for example, it was not enough that Richard Nassau described "headaches, nausea, dizziness to the point where you can almost taste it that it's so strong." Nor was it sufficient to point out that the dry cleaner who lived below him had been "tagged"—closed down by the city's health department for several months until he repaired his equipment. Dong Hwang of the Korean Cleaners Association of New York still had plenty of questions. "Did you go see a doctor or anything? What did you do about those problems you had? Is there any record of your sickness to prove that? I am not asking you as a question of integrity, but just as a matter of principle. Is there any verification of your being sick other than your statement?"

"Just my word," Nassau replied, "and my being here telling you so."[24]

Another witness, Tom Langner, a retired professor of psychology at Columbia University, also was subjected to such interrogation. The shop downstairs from the Langners used sealed, filtered equipment that supposedly protects cleaners and the public from dangerous fumes. Still, the sickening sweet smell of perc filled the Langners' apartment four stories above when there was a leak, a spill, a perc delivery, or, one time they knew of, when perc-soaked rags used for spot cleaning were piled in open bins.

Langner called the health department, but with only three environmental inspectors for all of New York City, they always arrived weeks after the smells disappeared. Only once, in spring 1993, when an inspector arrived on the day of a spill so severe it burned his nose, was the cleaner cited for a health violation. There was a three-month hiatus from perc problems, but in the fall the fumes began again. Now, the Langners were especially worried because they were expecting a baby girl in November.

Giving up on the public health authorities, the Langners bought perc-monitoring equipment of their own. One day, the detection tube turned pink up to the 25 parts per million mark. That was 1,700 times the limit that New York state authorities had set as a safety standard for apartment buildings. Another morning, they took a reading that showed a level of one part per million—nearly 70 times the state safety standard.

Not many residents who suspect chemical contamination in their homes can buy expensive detection equipment or train themselves to use it. Fewer still dare to come forward with their stories and face the barrage of cross-examination from industry lawyers and lobbyists that met Langner and others at the EPA hearing.

How did he know that smell came from perc and not some other chemical? Langner was asked this question in two different ways by officials of the Neighborhood Cleaners Association, and his answer was the same: The monitor he bought tested only for perc.

How was he sure that he was using the monitor correctly? Why did the city health officials not shut down the cleaner? Were any other measurements taken by others to verify his findings, asked Steve Risotto of the chemical-industry-financed Center for Emissions Control. Did other residents have problems? Why had they not come forward? And finally, how could he and his wife have smelled perc, when "the odor threshold," Risotto said, was 50 parts per million—double the highest reading recorded in the Langner apartment?

At this point in the EPA hearing, Judy Schreiber of the New York Health Department interrupted the questioning. Since 1991, when she began taking readings in apartment buildings and got very high levels of perc in units above dry-cleaning establishments, she had been trying to get better state and federal regulation of this obvious health hazard. She had found apartments as high as 12 stories above dry cleaners with perc levels as high as 60 times the state safety standard.

"I think it's very well recognized that there is a great deal of variability in odor thresholds, and some people are more sensitive and can detect an odor with their nose at a much more sensitive level than others," Schreiber said. "And I would not question your wife's saying that she can smell it. I would tend to believe her, especially given the circumstances where we have documented high levels in your apartment."

Risotto says that the problems in New York City stemmed from dry cleaners using old, poorly maintained equipment in an unsafe manner. "It was not endemic," he says. "It's in everyone's interest to reduce [perc emissions]. But in dry cleaning, the technology is good enough to drastically reduce emissions and exposures. The risk can be managed. The equipment is there."

But Consumers Union of the United States, which publishes *Consumer Reports* magazine, disagrees. In October 1995, it released a study of 29 apartments in Manhattan and Brooklyn in buildings with 12 dry-cleaning establishments that used modern, unvented equipment. Perc levels in all but five of the apartments exceeded the state safety standard. Perc levels in eight apartments were at least ten times higher than the safety standard, and the perc level in one apartment was 250 times the state standard. Consumers Union said that New York City's building codes should be changed to prohibit dry-cleaning establishments from operating in residential buildings, and it recommended that people who live near them test the air for perc.

A more dramatic illustration of how difficult it is to manage perc and its risks bubbled to the surface in suburban Washington, D.C. No property is more valuable than the few large plots of land that still sit vacant and waiting, not too far, yet not too close, to the nation's most important cities. In Silver Spring, Maryland, a group of real-estate developers who called themselves Westfarm Associates had every reason to believe they were sitting on a gold mine. Their property was along Tech Road, which, as the name would suggest, was part of the "technology corridor" that Maryland politicians hoped to carve in a grand suburban spur north of Washington.

But for Westfarm, the dream would be dissolved by a chemical. The land was ruined by perchloroethylene, and the way it was handled by the people who claim to know how to handle it best. It is a story that, more than anything else, puts the lie to claims that perc—although dangerous—can be handled safely. And it is a story of how chemical companies exact a cost that spreads gradually through society.

The Westfarm group learned of the pollution problem as they readied to sell their prize parcel of land in 1991 for $1.8 million to State Farm Insurance Company, which hoped to use it for a national claims-processing center. State Farm had insisted on a thorough environmental investigation of the property as a matter of course; the entire insurance industry is wary of toxic pollution, having been hit with more bills to clean up abandoned waste sites than any other single sector of the economy—courtesy of their manufacturer policyholders.

The tests showed that the site was soaked with perchloroethylene.

Engineers have a simple method of tracing the path of pollution, called triangulation. They dig three wells. The well most deeply saturated with the pollutant is like the point of an arrow stretched out from the other two wells, aimed at the pollutant's source. At the Westfarm site, the arrow pointed squarely at the offices of the International Fabricare Institute.

Officials of the institute set to work to shift the blame, using the same arguments they have employed to block government regulation of dry cleaners. First, they scoffed at Westfarm's suggestion that the perc came from the institute. True enough, they used perc in their clothing and chemical analyses for dry cleaners, but there were no leaks. They pointed out that perc has uses other than dry cleaning. The last time a dry cleaner was accused of pollution, the source was traced to an airplane hangar where perc had been used for metal degreasing, they said. There was no history, however, of airfields in Silver Spring.[25]

When Westfarm filed suit against the International Fabricare Institute, it made public many of the trade association's internal records, which more than explained how the perc might have escaped from its careful handlers. "Newer and larger patches of dead grass are apparent, strongly suggesting water containing perc is being thrown out the door," said one interoffice memo sent by Norman Oehkne, the institute's technical director.[26]

Earlier that year, Oehkne had sent another memo on the mishandling of perc by the people who train dry cleaners. Some staff members seemed to be suggesting handling fragile garments, such as sanded silk, by dipping them into a bucket of perc and drying them in the air. "Hand cleaning, etc., is not an environmentally safe method to be used, especially with perc," Oehkne said in another memo. "These procedures could result in a heavy concentration for both the operator and other personnel in the plant. Please do not suggest any procedure, like the above, that is in any way going to impose any problems."[27]

To the outside world, the institute presented a different face. On May 1, 1989, for example, Jon Meijer, a vice president, filled out a required annual report to the state of Maryland suggesting that it strictly controlled its hazardous waste. "IFI strips most spent filters and they are sent for recycling," he said in the report. "Steam stripping removes approximately one gallon of solvent per cartridge filter. IFI also uses steam stripping to reduce concentrations of perchloroethylene in still residues."[28]

Just four days later, on May 5, 1989, Meijer sent a memo to Bill Fisher, the IFI's president and chief lobbyist, about the multiple, growing problems with the very hazardous wastes that he had told the state of Maryland were well in hand. The ballyhooed cartridge stripper, for example, was "out of commission," and not for just a few days, weeks, or even months.[29]

"Over the past couple of years, we have made several attempts to get this unit back into operating condition, both for treatment of our wastes and for demonstration purposes with the classes," Meijer wrote. "In fact, I believe that I am the only person who knows how to operate the unit."

In court, the IFI's attorney, Duane Siler of Patton, Boggs & Blow (the same big Washington law firm that represented Dow Chemical and other makers of perchloroethylene), presented an explanation for this damning memo. Under his patient questioning, Meijer said that the IFI's dry-cleaning machines had automatic stripping capabilities. The cartridges did not need to be stripped a second time.

But uncontradicted in the courtroom was the comment he made in the same memo to Fisher about the perc-saturated filters and muck, the so-called still residues. They appeared to be 50 to 60 percent perc, grossly overpolluted compared to what the levels the IFI recommends to dry cleaners: 1 to 2 percent.

More serious than any of these problems, however, was the discovery made by Ivan J. Andrasik, a Czech immigrant who had worked at the IFI for 23 years. After the Westfarm suit was filed, he remembered the solvent analysis that the IFI had conducted free for Dow customers. With every test, 10 milliliters of perc were dumped down the drain.

"I started to think about it because we are trying to be environmentally conscious," he said in pretrial questioning by Westfarm's attorney.

"Then it struck to my horror that we actually are dumping some perc into sink....Why don't we just save it like we do...with everything what contains perchloroethylene and give it to Safety Kleen [a hazardous waste hauler]....Now that was something—an idea which was long time overdue."[30]

From 1969 to 1991, the IFI had dumped at least 150 gallons of perc-contaminated "separator" water down the drain each and every year. And it continued to do it at least a year after telling its members in a May 1990 bulletin that the separator water should not go down the sink.[31]

At trial, the IFI's lawyer, Duane Siler, tried to minimize the pollution problems. The residue of dry cleaning is "not an obnoxious black substance," he said. "It looks more like honey." All of the IFI's waste was put in containers "except for a little fingerful going down the drain," he said. "Was IFI somehow negligent? There was nothing unreasonable about PCE, or perc. It's the most common dry cleaning fluid, and it offers certain environmental benefits." (That is, compared to the gasoline products used in some places, it does not blow up.)

On the witness stand, Fisher expanded on this theme: "Just like ammonia or gasoline, one does not drink perc. Therefore, if handled with care, its general effects are minimal."

Leonard Breitstein, a Bethesda, Maryland, engineer who testified on behalf of the Westfarm real estate group, had a different opinion. "There are both carcinogenic and non-carcinogenic effects of perc," he said. "It affects the primary central nervous system. It causes symptoms like dizziness, headaches, nausea, and worse. There are also adverse effects on the liver and kidneys."

"Can these risks be eliminated by the exercise of reasonable care?" asked Westfarm's lawyer, Jeffrey Johnson.

"My opinion is that they can be reduced, but they cannot be eliminated," Breitstein replied. "You are always going to have spills and leaks and human errors and things that cause problems, I mean as hard as you try to eliminate the problems. And I think this is typified by the problems at IFI, where a number of spills and leaks have in fact been documented by their own records."

"With respect to IFI, these people are not bad people," Johnson said in his closing statement to the jury. "Unfortunately, however, they were contemptuously familiar with a chemical, perchloroethylene."

What happened next in the case is an example of how everyone—in one way or another—pays for the adverse effects of toxic chemicals. Before trial, as the evidence against the institute began to build, its lawyers at Patton, Boggs sued the local sewer system, whose pipes the IFI had polluted with perc.

Their argument was that the institute's perc would not have seeped into the soil if the sewer system had not been so leaky. Westfarm, in an effort to recover funds from any possible money sources, joined the IFI in blaming the sewer system, at least in part, for the pollution. Westfarm and the IFI even took a videotape from inside the sewer lines. It showed the kinds of cracks, uneven joints, and flaws that can be found in every sewer system in the United States.

The jury determined that the sewer lines were indeed faulty and returned the verdict that the clean-up costs at the site would have to be shared with the residents of suburban Washington, who pay monthly bills for service from the same sewer system. Attorneys for the International Fabricare Institute and the sewer authority were still wrangling over how to split the costs in early 1996, generating legal bills that also will be footed by the sewer system's ratepayers.

7

The PR Juggernaut

The report had all the ingredients of a blockbuster. The Environmental Working Group, a Washington-based environmental organization known for its success in attracting press coverage, was poised in the fall of 1994 to release a startling study on the contamination of midwestern water supplies by weed-killers that wash off farmland. The report would disclose that more than 14 million people—including 65,000 infants who are fed reconstituted formula—regularly drink water contaminated with five major weed-killers, including atrazine and alachlor.

If you do not remember the report, *Tap Water Blues: Herbicides in Drinking Water,* or a massive outcry or response by government officials, it may be because the nation's biggest pesticide manufacturers handled the study as a public-relations crisis before it could become a public crisis.

Quarterbacking the industry's response was Adele Logan, who at the time was the senior director of communications for the American Crop Protection Association, the industry's chief trade group. The association had heard about *Tap Water Blues* through "press and political contacts" a few weeks before the report was to be issued, according to a brochure touting Logan's abilities produced by Strat@comm, the Washington-based public relations firm where she now works. Logan knew that it was the most ambitious effort yet by the Environmental Working Group, whose earlier reports about the dangers of pesticides had attracted

173

widespread attention. So she sprang into action. She immediately designed and distributed pro-herbicide mailings to journalists around the country and set up interviews with "experts" chosen by the industry. She alerted key state officials and farm organizations about the report's imminent release and conducted a preemptive media campaign.

"It's PR 101," Logan says, recalling her efforts on behalf of the pesticide industry. "Whenever a group is going to release a report, if you can you try to get that report. You try to find out about it and you handle the media inquiries as they come up. Some of that involves third-party experts. The important thing is to get the information in the hands of the people who need it and to do it in as concise a manner as possible. Be it a reporter or an allied group or a member company, what you're trying to do is make sure they have the information they need. The industry realizes that it's got a responsibility to tell its side of the story."

Just how well her preemptive strike worked depends on who you ask. Richard Wiles of the Environmental Working Group says that *Tap Water Blues* was mentioned in more than 1,400 news articles or broadcast reports, including segments on two network news shows and on ABC's *Good Morning America*. "It was unquestionably a huge media success from our standpoint," Wiles says. But Logan's advance work clearly had its intended effect. Interviews that she arranged were featured on three network news broadcasts and in more than 110 news articles, according to the Strat@comm brochure. The bottom line, as the brochure states baldly, is that "because of the dearth of negative publicity, ACPA's member companies avoided additional regulation of their products."

For decades, the chemical industry did not have to worry about public relations. Americans accepted without question the idea that man-made chemicals benefited companies and consumers alike. That changed dramatically on June 16, 1962, when the *New Yorker* magazine published the first of three articles by Rachel Carson, a biologist and nature writer, who argued passionately and compellingly that pesticides were dangerous and overused. The articles were an immediate sensation, and two months later they were published in the instant bestseller *Silent Spring*, the best-known environmental book ever published.

Carson's devastating critique plunged the industry into its first full-blown public relations crisis, and manufacturers, initially caught unprepared, quickly regrouped and responded with guns blazing. E. Bruce Harrison, a young public relations executive, was put in charge of managing the industry's response to *Silent Spring*. He engineered a series of attacks on Carson delivered by industry-friendly scientists in press interviews and in pamphlets and negative book reviews sent to thousands of prominent Americans.[1] Monsanto distributed more than 35,000 copies of an acerbic rebuttal in its company magazine. Called "The Desolate Year," the article described a pesticide-free world in which insects ran rampant, spreading famine and disease. The anti-Carson campaign was so effective that *PR News*, a trade publication, would later enshrine it in a collection of standout public-relations efforts. The industry's "brilliant PR performance," *PR News* said, "...has raised high a big 'Stop and Think' sign which has slowed hysterical, dangerous extremism."[2]

Because of the impassioned response to *Silent Spring*, the nation's chemical industry would never again take public support for granted. It has poured billions of dollars into advertising and public relations campaigns as part of an all-encompassing strategy that has, for the most part, overwhelmed the efforts of independent scientists to present a balanced picture of the risks and benefits of man-made chemicals. Harrison's counterattack against *Silent Spring* established a pattern that manufacturers would follow again and again: using the overwhelming power of public relations to blunt the impact of negative publicity— "crisis management," in the lexicon of the fast-growing environmental public relations industry.

Adele Logan is a protégée of the master; before joining the pesticide manufacturers' group, which was then known as the National Agricultural Chemicals Association, she had worked at E. Bruce Harrison Company, the first public relations firm to focus entirely on environmental issues. (Strat@comm was founded by two other Harrison alumni, and the Harrison firm was itself acquired by mega-agency Ruder Finn at the end of 1995.) At Harrison, Logan specialized in tough sells: tobacco, auto pollution, and pesticides. She continued that role at the NACA, organizing meetings with newspaper editorial boards and teaching industry officials how to talk to the press. During her tenure there, the NACA changed its name to the more benevolent-sounding American Crop Protection Association.

The contrast between Harrison's improvised, after-the-fact response to *Silent Spring* and Logan's preemptive strike against *Tap Water Blues* 32 years later is a telling one. The industry has honed its public relations skills so effectively that it will never again be caught completely unaware.

Logan's aggressive efforts are typical of the chemical industry's willingness to do—and to spend—whatever it takes to blunt negative publicity.

"If they need to dump more money to solve the problem, money's no object," says John Stauber, the executive director of the Center for Media & Democracy, a research and advocacy group in Madison, Wisconsin. Stauber is the editor of *PR Watch*, a newsletter that spotlights misleading corporate PR campaigns, and the author of the book *Toxic Sludge Is Good for You*. "There's simply no roof on what they can spend," he says. "If they get in a jam, literally overnight they can double or triple their resources. Really you're talking about crisis management, and the chemical industry is in a constant state of crisis."

Sometimes chemical manufacturers work quickly and effectively to squelch immediate threats to their products. At other times, the crisis is not so obvious, as when warm Sunday-morning television ads bathe viewers in confidence and good feelings about the chemical giants. But whether it is working on a targeted or blanket public relations project, the industry's approach has become increasingly subtle and indirect, as it tries to assuage the public's fears about toxic products. Its most effective campaigns enlist workers, customers, and other ordinary citizens to court the public and the regulators, while the manufacturers, like Cyrano, whisper the eloquent pitch from the background.

Polishing Its Public Image

Chemical manufacturers have spent billions of dollars on advertising campaigns designed to make people forget that they manufacture chemicals.

It was not always this way. DuPont used to promise "better chemicals for better living." In popular color magazines of the 1940s and 1950s, like *Better Homes and Gardens*, the Lead Industries Association would show the smeared, ruddy face of a miner above the caption, "The lead we're mining here is what puts gumption in paint." That was years before the public knew—though the industry had known it for a long time—that lead also put poison in the paint.[3]

Today, chemical manufacturers have abandoned straightforward defenses of synthetic chemicals in favor of soft-focus campaigns that feature corporate logos and slogans superimposed on pictures of romping children, smiling graduates, swaying cornfields, salt-of-the-earth farmers, and other Norman Rockwell–style images. DuPont has dropped the c-word and changed its slogan to "Better things for better living."

It was not only Rachel Carson and the environmentalism of the 1960s that spurred this change. Anti-war activists of the same era singled out Dow Chemical for protest because of its development and production of napalm, the flaming chemical weapon that seared the skin of soldiers and civilians alike in Indochina. By the 1970s, there were widespread suspicions that U.S. troops in the Vietnam War had also been ravaged by another Dow product: the tropical forest defoliant Agent Orange. In 1984, after settling numerous lawsuits brought by veterans, Dow commissioned a poll that showed the majority of the public viewed the corporation as "insular" and "arrogant." The newly skeptical public no longer saw chemicals as benign miracle substances but as killers.[4]

In the mid-1980s, Dow launched a campaign that would be a vanguard of the new chemical advertising, by turning on its head the image of moral outrage on campus that had become so associated with the company. The theme "Dow lets you do great things" was the corporation's first "nonproduct" advertising on national television. In one commercial, a young woman reflects on her life on graduation day: "When I was little, I never understood what mom meant when she said, 'Clean your plate, Cindy, there are places where kids are starving.'" She glances across the sea of fellow seniors in caps and gowns to catch her mother's eye. "Well, Mom, now I am old enough to understand, and I love you and daddy for teaching me to care." Then Cindy says, "That's why I'm eager to walk into a Dow laboratory, to work on new ways to grow more and better grain, for all the kids who need it so desperately." As she receives her diploma and mouths the words, "I can't wait!" her parents, the dean, and an older woman professor, who appears ready to burst into joyful tears, beam.

After the "Feed the World" ad and others with similar imagery, a 1986 opinion survey showed a revolution in public perceptions of Dow since the company's temperature check just two years earlier. Customers, scientists, and employees showed highly favorable attitudes toward Dow, and there was a striking 60.5 percent gain in favorable media opinion.[5] Dow's new PR recipe proved as successful as any formula developed in its labs.

In *Going Green*, a book describing his work that was published in 1994, Harrison says that he always starts his coaching of business executives by handing them a small card that fits comfortably in the palm of the hand. It says, "We are the good guys!" He tells them to look at it whenever they challenge on an environmental issue.

Georgia-Pacific burnishes its good-guy reputation in some of its brochures for plywood paneling, a product that contains formaldehyde. Passing over the issue of toxics, it bills itself as a corporate conservationist. "As managers of more than six million acres of forest land in North America, we consider it our business to be good stewards of the land," the brochure says. "We replant or naturally regenerate more than 130 million seedlings on company land."

Monsanto leapt at the chance to show itself to be a corporate good guy in the aftermath of the record floods of the Mississippi River and its tributaries in the summer of 1993. The company and one of its public relations consultants, Drake & Company, designed a strategy that would not only raise millions of dollars for the flood victims but also generate huge publicity for the company and win the goodwill of midwesterners, especially farmers. The company called a press conference, attended by powerful hometown Representative Richard Gephardt, then the House majority leader, at which Monsanto presented a $1 million check to the American Red Cross and promised to match up to $1 million more in contributions from other groups. With Drake publicizing the event, the donation was picked up by CNN's *Moneyline* program and ABC's *World News Tonight*, as well as newspapers and television stations throughout the Midwest. Elizabeth Dole, the president of the American Red Cross, then touted the contribution during a one-hour telethon at which an additional $1.2 million was raised, generating another round of favorable stories.[6]

Another PR firm, Dorf & Stanton, succeeded in getting major coverage of Monsanto's contribution in farm magazines and newspapers. The campaign was such a public relations success that even Monsanto's chief competitor, Ciba-Geigy, donated $500,000. In all, more than $5 million was raised in the campaign, including $2.3 million from Monsanto, according to company estimates.

The Mississippi floods were "an intensely personal thing for the employees of the company in St. Louis," says Monsanto spokesman Dan Holman. "To watch a levee burst on live TV and flood an area where we've driven through, and through farms where we've gone and bought pumpkins at Halloween, and Christmas trees at Christmas, you see that and you can't help but be touched by it."

Most of the stories overlooked another "gift" from Monsanto, Ciba-Geigy, and other herbicide manufacturers that was engendered by the Mississippi River floods: record-high quantities of herbicides in the river that millions use as a source of drinking water. In July 1993, atrazine was flowing in the Mississippi past Thebes, Illinois, at a rate of more than 12,000 pounds per day, according to the U.S. Geological Survey. A year earlier, the rate at Thebes was less than 3,000 pounds per day. Of course, the amount of water flowing past Thebes was also dramatically higher during the floods. But scientists who had expected that the torrential river flow would dilute the impact of the weed-killers were surprised to find that the herbicide concentrations did not decrease at all.

"The concentrations were as high as they had been during the previous years when [water] flows were 50 percent less," Donald Goolsby, the chief of the U.S. Geological Survey's midcontinent herbicide program and a leading expert on herbicides in drinking water, recalls. "We were surprised. Given the high flows, we expected some dilution, but we didn't see it. When we put out a report and a press release, the industry people weren't happy about it."

But the government report got relatively little attention amid the powerful stories and pictures of families flooded out of their homes—and the hometown corporation that was reaching out to help them. The Public Relations Society of America was impressed enough with Monsanto's campaign that it presented the company and its public relations firm with a 1994 Silver Anvil Award. "News coverage ensured that residents of the Midwest were aware of the fund and Monsanto's commitment," the society noted. "Editorials in agricultural publications praised the effort." Another group, the Agricultural Relations Council, gave Drake its top public relations prize, the Founders Award. The American Red Cross named Monsanto its "Philanthropist of the Year."

Georgia-Pacific got a similar media boost when it pledged to protect the habitat of the red-cockaded woodpecker in its logging operations, a pact that Interior Secretary Bruce Babbitt obtained in the first months of the Clinton administration. After all the battles between timber companies and the spotted owl in the Pacific Northwest, the deal, as the *Christian Science Monitor* put it, was "one small harbinger of peace."

The company was on the front page of the *New York Times* and every major newspaper as the standard-bearer for a new era in corporate concern.

No mention was made in the coverage that, less than four months earlier, Georgia-Pacific had been branded one of the nation's eight "biggest polluters" by the New York City–based Council on Economic Priorities. Its chemicals had spilled and caused massive fish kills in Mississippi. In Maine, the company had dumped six million gallons of wastewater into the St. Croix River and been slapped with a record state fine. The EPA fined Georgia-Pacific seven times from 1977 to 1990; only DuPont and USX Corporation had been fined more often.[7]

When chemical manufacturers are not working on soft-focus feel-good campaigns, they often are spinning some variation on the theme that Monsanto presented in its old ads: "Without chemicals, life itself would be impossible." Man-made chemicals, the commercials told us, are our hope for the future; all of us should fear what might happen should they be taken off the market.

When the EPA was considering banning alachlor, for instance, Monsanto launched an all-out effort to galvanize the nation's farmers. George Fuller, the director of product registration and regulatory affairs for the company's agricultural division, explains its strategy this way: "There's not a lot of corn growing in Washington, D.C. There's not a lot of awareness of exactly what it takes to grow corn and what are the issues that are facing corn farmers. What we wanted to do was create that awareness and have that be a part of the exchange of ideas."

In its direct appeals to farmers, the company's message was considerably blunter. Monsanto's Ed Rieker used the language of fear to urge farmers to flood the agency with letters of opposition. In formulating these letters of protest, he told the farmers, they should remember that "in the United States, housewives have never purchased wormy cabbage or buggy cereal" and that "ignorance by the non-agricultural community concerning pesticides, especially in relation to overall public health, is a very frightening affair."[8]

Often, the message of fear is delivered with subtlety. A 1985 television commercial for Dual, Ciba-Geigy's metolachlor herbicide, is typical of the genre. A city dweller is pictured wandering through an abandoned farm as the announcer extols the abilities of the American farmer as veterinarian, business executive, agronomist, and engineer.

What would happen, the announcer asks, if all of America's farmers suddenly disappeared? "Don't take farmers for granted," the announcer finishes, as the words "Long-Lasting Dual" appear on the screen.

Formaldehyde manufacturers took a different tack in the early 1980s, when consumers first began to raise questions about fumes from formaldehyde and the studies showing its carcinogenicity. They called formaldehyde "a building block of our society" in informational pamphlets distributed to public health and environmental authorities.[9]

Formaldehyde has a simple molecular structure: a couple of hydrogens and an oxygen—just like water—joined with a single carbon atom. The "building block" campaign sought to exploit the average American's limited knowledge of chemistry.

The Formaldehyde Institute also sought to evoke fear of life without formaldehyde. The chemical, it said, was "an indispensable ingredient in the production of thousands of industrial and consumer products"— products that account for 8 percent of the gross national product and involve 1.4 million American workers in 45,000 U.S. factories.

The literature went on to say: "Obviously, significant restriction of formaldehyde would have some very serious economic implications. It would also touch consumers deeply and directly.... [Without formaldehyde,] the clothes we wear, the cars we drive, and the houses in which we live would be changed significantly. Without formaldehyde, the sawdust and wood chips now bonded into particleboard would revert to the waste fuel they once were, instead of being transformed into valuable building materials. Some products would become prohibitively expensive and probably disappear from the market entirely."

Front Organizations

Georgia-Pacific, DuPont, and the other heavyweights in the formaldehyde industry had a big problem on their hands by 1977. Evidence was beginning to surface that the chemical was carcinogenic, and the companies were especially concerned that a key constituency—mobile-home dealers—was on the brink of turning against them.

Within two years, however, the big formaldehyde manufacturers had managed to recruit the dealers, and just about everyone else involved in the sale of their products, to be in the front lines of lobbying to keep formaldehyde on the market. They had turned the mobile-home dealers into the most effective public relations weapon a chemical manufacturer can have: a "grass-roots" constituency.

The most sophisticated—and effective—PR campaigns, after all, feature heartfelt appeals from small-business owners, farmers, and other "average citizens." They are often combined with reassuring messages from other industry surrogates: purportedly independent research and advocacy organizations. Formaldehyde manufacturers were able to orchestrate just such a two-pronged campaign through the creation of the Formaldehyde Institute.

At the time, the foul odor of formaldehyde was becoming a big issue among mobile-home owners and dealers, particularly in Texas and other southern states where heat and humidity made particleboard emit higher levels of formaldehyde gas. And mobile homes typically have a higher proportion of pressed wood and particleboard than other dwellings.

R. Bruce Walters, the chairman of the Texas Manufactured Housing Association, wrote a letter to 15 manufacturers of formaldehyde-impregnated paneling, flooring, and decking, asking for help. "When the houses are closed up on the dealer lots, the inside air becomes saturated with a very strong chemical odor," Walters wrote. "We do not know, for sure, which of the materials is to blame, but we do know that the plywood panels cause a significant amount of the problem. It is so strong that it causes tears and drives prospective buyers back outside."

Walters said that dealers had tried cut apples, charcoal, and several other home remedies to neutralize the odor, without success. "Would you please ask your chemists what we can use that will counteract and deodorize this new panel and/or other components?" he asked. "We do not like to make our customers cry. We want to see them smile and I think you do too!"[10]

The wood-products manufacturers were, in fact, just as worried as Walters and his members. The vast majority of manufacturers simply purchased urea-formaldehyde resin from the chemical companies and mixed it with wood chips. Leaders from 21 wood-products companies (organized as the National Particleboard Association) met in Houston that fall to draft a letter to the chemical companies "to remind manufacturers that the source of the problem is their material."[11]

Only one products manufacturer at the meeting also manufactured formaldehyde: Georgia-Pacific, the biggest company in the industry. The executives of Georgia-Pacific and the other major formaldehyde manufacturers—Borden, Celanese, and DuPont—would spend the next couple of years working on a strategy to deal with this rebellion by customers. The idea was not to battle them as adversaries but to turn them into allies. This was the birth of the Formaldehyde Institute.

On February 21, 1979, Clifford T. "Kip" Howlett, Georgia-Pacific's director of environmental issues, wrote a memo to his boss, Thomas F. Mitchell, the company's vice president for government affairs, after a meeting in Dallas that was aimed at boosting membership in the institute. Howlett made clear that James Ramey, a Celanese executive who had been recruiting members for the institute, was disappointed at the reluctance of the big wood-products manufacturers to join.

"The FI is off to a rocky start," Howlett wrote. He added that there have been "frank and open discussions" as well as "a good deal of posturing." He went on to say: "Unfortunately, certain doubts and misgivings were expressed mostly by the Hardwood/Plywood Manufacturers Association. All of the smaller users were very enthusiastic about the program. All of this, I think, frustrated James Ramey who...unfortunately fantasized that a quick and positive response would be made by all, and that just didn't happen."

Within the next two years, however, the Formaldehyde Institute, under fire from federal and state regulators, would become an increasingly cohesive and considerable force. Through an all-out campaign of lobbying, lawsuits, the sponsoring of scientific studies, and public relations, the institute would beat back every government effort to intervene, even as the evidence of health risks from formaldehyde mounted. And instead of lobbying the chemical companies to change their resin formulas, the wood-products manufacturers joined them front and center.

Proof comes in the minutes of an April 1980 meeting of the board of directors of the Texas Manufactured Housing Association—the same group that, under Walters's leadership three years earlier, had complained about the products that were making its customers cry. Now, the association was a member of the Formaldehyde Institute, and its director, Will Ehrle, gave advice on what sellers should say to the question, "What is that smell?" Ehrle, according to the minutes, said "to respond that it is a substance used in various building materials, and not to use the word formaldehyde.'"

Stephen K. Jackson, Georgia Pacific's vice president for advertising and public relations, wrote a memo to his file on February 23, 1982, summarizing the company's plans for the Formaldehyde Institute, as well as the benefits it would reap. Georgia-Pacific, he wrote, would support a blue-ribbon panel to analyze the existing data on formaldehyde— "necessary for potential legal and legislative action as well as public-relations opportunities." It would also encourage the institute to pull together a summary of formaldehyde's benefits. "The above," Jackson wrote, "is designed to keep Georgia-Pacific out of the limelight, yet as a psychological force within the Institute."[12]

The Formaldehyde Institute is gone, but the companies that manufacture formaldehyde still prefer to stay out of the limelight rather than discuss its potential hazards. Only Hoechst Celanese agreed to speak about the history of formaldehyde and its regulation for this book through Jerry Dunn, the company's vice president for environmental, safety, and health affairs. DuPont had no one to talk in detail about formaldehyde because, its spokeswoman said, it no longer sells the chemical on the market, having switched in the early 1990s to producing only enough for use in its own chemical products. Georgia-Pacific, the number-one maker of formaldehyde wood products, would not provide anyone to talk about the issue.

Borden, the nation's biggest producer of formaldehyde resins, initially referred questions about formaldehyde's history and regulation to its outside attorneys. Borden's director of corporate communications, Nick Iammartino, said that one of those attorneys was Kathy Rhyne of the Washington office of King & Spalding, an Atlanta-based firm, who he said was representing the Formaldehyde Epidemiology and Toxicology Working Group. When interviewed, Rhyne said that she preferred to be identified as someone who did work for "resin manufacturers," not for any formal working group. "My clients are gun-shy about becoming a target for litigation," she said. They would consider such litigation frivolous, she added, because they do not believe that their products pose any hazard.

Few industry-sponsored organizations have literally been put out of business by the high costs of obtaining insurance coverage for lawsuits, as the Formaldehyde Institute was. And the advantages of presenting a unified front remain great enough that groups like it have become a staple of public relations campaigns in the chemical industry.

Consider the Center for Emissions Control, which in May 1994 gave many journalists their first notice that Greenpeace, the well-known environmental organization, was about to release its study of perchloroethylene. Three days before the planned announcement, it sent a package to reporters that purported "to help familiarize you with the issue and provide some background."[13] The Center for Emissions Control has a more appealing name than its sister organization, the Halogenated Solvents Industry Alliance, which is located in the same offices. (Chemicals that are "halogenated" are those treated with chlorine.) Next door is the parent organization of both outfits, the Chlorine Institute, an alliance of chlorine manufacturers and users.

The Iowa-based Council for Agricultural Science and Technology touts itself as an "unbiased" source of scientific information for reporters and politicians, but it is a creation of agribusiness companies, including pesticide manufacturers. In a directory published in a public relations newsletter, the multinational PR firm of Shandwick Public Affairs boasts that its achievements include "establishing a non-profit research organization—the Council on [sic] Agricultural Science and Technology—as the source for public policy-makers and news media on environmental issues."[14]

The council's financial statements show that agribusiness interests are its single largest source of funds. They accounted for at least $273,953 of the council's $620,553 in revenues in 1994, according to its annual report for that year. Many of its publications have been aimed at countering studies by environmental organizations on such topics as groundwater contamination and the use of genetic engineering to develop herbicide-resistant crops. The council counts as one of its "highlights" for 1994 its "instrumentality in killing a major television news media story before it was aired, when [a] reporter realized that the science provided negated his story."[15] The never-aired story was about antibiotic residues in cows' milk.

The most effective surrogates of all, however, are the chemical industry's covertly orchestrated "grass-roots" coalitions. To recruit foot-soldiers for these coalitions, manufacturers pound home the message that using chemicals is the only way to prosper. Farmers, for example, are told again and again that pesticides are as essential to their prosperity as rain and sun. Chemical companies are the dominant advertisers in magazines such as *Farm Journal* and *Farm Chemicals*, where there is virtually no distinction between advertisements and "news" articles.

Ciba-Geigy, for example, was the sole advertiser in the summer 1991 issue of *Farm Chemicals*, in which an article about how to lobby Capitol Hill lawmakers was sandwiched between four-color advertisements that offered "straight talk" about atrazine and other Ciba-Geigy chemicals. An editorial in the February 1995 issue of *Farm Journal* not only aggressively backed atrazine and urged farmers to write the EPA on the chemical's behalf, it even included addresses and tips on what farmers should say in their letters—just like the Ciba-Geigy advertising that appeared elsewhere in the publication.[16]

Farm broadcasters are an equally popular target. Chemical manufacturers eagerly provide stories—complete with their own videotape and company-friendly scripts—to television stations. Ciba-Geigy bankrolls the National Association of Farm Broadcasters/Ciba Plant Protection Farm Broadcasting Award Program, which confers the top prizes for television news about agriculture. Because four regional prizes and a nationwide award are handed out each year, many farm broadcasters eventually win.[17]

Even more significantly, virtually all of the advice that farmers get comes from sources aligned with the pesticide industry. The salaries of county cooperative extension agents are paid by taxpayers, but the land grant universities for which the agents work are financially dependent on research grants from chemical manufacturers. And the chief "educators" of most farmers are not county extension agents but the distributors and roving salespeople who work for Ciba-Geigy, DuPont, Monsanto, and other major chemical companies. Monsanto has even organized "educational seminars" in which farmers, over lunch, listen to employees of the company talk about how best to control weeds. (Hint: It involves Monsanto products.)

Manufacturers are not nearly so enthusiastic when it comes to telling farmers about the hazards associated with their products. Ordered by the EPA in 1985 to produce an alachlor safety video for farmers, Monsanto came up with a script that the agency said was misleading, partly because it soft-pedaled the weed-killer's risks. However, by the time the EPA lodged its criticism, Monsanto had already produced the video, and thousands of farmers had seen it. The company resisted the agency's proposed changes in the script and ultimately produced four drafts before the two sides agreed to compromise language, according to EPA records obtained through the Freedom of Information Act.[18] Monsanto says that the EPA was being "overly conservative." Monsanto's George Fuller puts it this way: "Why would we put out a video that we didn't believe in? It shouldn't be a shock to understand that we did not have the same approach initially that they did, but we were able to come to an agreement."

Ciba-Geigy, on the other hand, simply dismissed similar criticism from the EPA about a video on atrazine safety that the company produced and showed to tens of thousands of farmers across the nation in the early 1990s. Unlike Monsanto's alachlor safety video, Ciba-Geigy's was not produced under orders from the EPA, and the company did not have to act on the agency's strongly worded critique. An EPA letter to Ciba-Geigy, obtained from the agency under the Freedom of Information Act, shows that the EPA was uncharacteristically blunt in its criticism of the video. The EPA working group that reviewed the Ciba-Geigy video used especially strong language in assessing the video, which was supposed to show farmers how to minimize atrazine contamination of drinking water. The working group "was concerned that the public-relations goal (there is no problem) got in the way of the behavioral change goal," Ann Sibold, the EPA's review manager, said in the letter to Ciba-Geigy. "Overall, the presentation of the big picture on ground and surface water contamination did not seem balanced."[19]

Asked about the episode, Ciba-Geigy executives say that there are two reasons the company did not change the video: Ciba-Geigy disagreed with the agency's comments, and it had already distributed the tape widely by the time the EPA issued its critique. "We had already used the video by the time we got the comments back from EPA," says Richard Feulner, Ciba-Geigy's director of regulatory affairs.

With pesticide manufacturers exercising virtually ironclad control over the information that reaches farmers, it is no wonder that many growers are ignorant of—or even hostile to—alternative farming methods that do not rely on the heavy use of chemicals. "There's this real fear kind of atmosphere, maybe it's even a little nastier than fear," says Rick Exner of the Iowa State University Cooperative Extension, who works on alternative agriculture programs. "Agribusiness has influence at all levels," says Rhonda Janke, an associate professor of agronomy at Kansas State University and the only alternative farming specialist on its faculty. "The farmers that I work with out here in Kansas, they'll talk with me about alternative agriculture and then go to their chemical-dealers meeting and have steak dinners. Their friends and neighbors are the people who come out and spray their fields. It's a way of life out here."

Dennis Keeney, the director of Iowa State University's Leopold Center for Sustainable Agriculture and a leading critic of chemical-based farming, puts it even more succinctly. "The chemical companies," he says, "are the driving force for agriculture."

A look at the dry-cleaning industry shows just how effective public-relations campaigns can be in turning product users into anti-regulatory crusaders. Federal occupational and environmental health officials have noted the extreme commitment that dry cleaners have for perchloroethylene, despite its cost, the disposal difficulties it creates, and the emergence of wet cleaning as a realistic alternative. Scientist Avima Ruder, who had studied cancer among employees of dry-cleaning establishments for the National Institute of Occupational Safety and Health, wondered why, so Ruder and her colleagues set out to do something that NIOSH had never done before. In early 1995, they began sitting down in small groups with dry cleaners to ask them what they knew about perc, the health risks it posed, and the possibility of alternatives.

In group after group, the results were the same: Without perc, the dry cleaners told the NIOSH scientists, they would be out of business. It is not surprising that cleaners believed that a solvent was the only thing that stood between them and bankruptcy; they have been bombarded with propaganda from perc manufacturers since the 1950s.

"Use of inappropriate risk characterizations to describe perc has the potential to ruin the owners of the 30,000 stores in our dry-cleaning industry," a column in the February 1995 issue of *Fabricare News*, the newsletter of the International Fabricare Institute, said. "Possible Cancer Scare," said the headline on the column by William E. Fisher, the institute's chief lobbyist.

Although the organization appears to have a diverse membership of mom-and-pop dry cleaners, in truth, it also is subsidized by the producers of perchloroethylene. Dow pays the IFI for conducting tests on solvent mixtures for dry cleaners that buy their perc from Dow. And the IFI uses the same powerhouse Washington lobbying firm, Patton Boggs, as the Halogenated Solvents Industry Alliance, the perc industry's trade group. The first large-scale dry-cleaning conference to deal with environmental issues was "Vision 2000," which the IFI sponsored in 1994. The panels were divided into those "outside" the industry (federal regulators, environmental groups, the news media) and those "inside" the industry (dry cleaners and the solvent makers). Not only were some of the regulators perturbed at the meeting's "us versus them" framework, but they also wondered why the big chemical companies were considered "insiders."

Another such organization is FLARE, an acronym for Fabricare Legislative and Regulatory Education, which includes IFI but is led by executives of R. R. Street & Company of Naperville, Illinois, which sells dry-cleaning equipment and distributes the perc made by Vulcan Chemical Company, a small rival to Dow and ICI.

"Cleaners and other businesses have used perc for many years, and its health and environmental effects have been very well studied," FLARE said in a mailing to dry cleaners in December 1994. Both statements are true, though the letter's description of the study results was misleading at best. The mailing went on to cite a 1991 statement by the EPA's Science Advisory Board that it saw "no compelling evidence of human cancer risk" for perc; the language had been included only at the insistence of the International Fabricare Institute's representative on the advisory board. What's more, two studies published in the previous year had provided compelling evidence of human cancer risk from perc.

In January 1995, *American Drycleaner* magazine published a 10-page article ("The Government, the Environment, and You") by Manfred Wentz, a lobbyist for R. R. Street. Warning of the EPA's possible action on perc, Wentz concluded his piece with a political rallying cry. "With the FLARE network," he wrote, "all dry cleaners have now a grass-roots organization in place at every political and regulatory level for responsible and effective political and regulatory participation, communication, and action." The message pounded into every dry cleaner is that those who push for regulation of perc are trying to put them out of business, while the industry-led organizations are their bulwark in the storm.

An account of one meeting in the February 1995 issue of *Drycleaners News*, the newsletter distributed by the Neighborhood Cleaners Association of New York, gives a glimpse of a "grass-roots" network in action. It reports how representatives of dry-cleaning equipment manufacturers, perc distributors and suppliers, and trade associations talked about organizing a "garment care truth squad" to fight Greenpeace, labor unions, and others that seek regulation of perc. It quotes Bill Seitz of the Neighborhood Cleaners Association as saying that he thought $1 million could be raised, with the help of such "deep pockets" as Dow and the Center for Emissions Control. And Dan Scharf of R. R. Street suggests that the groups go on the offensive against Bonnie Rice, the Greenpeace activist who had been working on perc. "Use a little muscle," Scharf is quoted as saying. "Call her a liar and get a mention on Donahue. To get media coverage you need confrontation."

Stamping Out Fires

Even the slickest image-building efforts cannot shield chemical manufacturers from public relations crises. It is, after all, an industry that deals in poisons. Bad press, ugly lawsuits, and regulatory wars are all part of the price of doing business—so much so that pesticide companies routinely spend far more on marketing and regulatory issues than on actually manufacturing their products. It is no wonder, then, that coping with public relations crises has been a central obsession of the industry ever since the publication of *Silent Spring* helped to launch E. Bruce Harrison's career.

In 1991, for example, Monsanto was so eager to blunt bad publicity over the use of bovine growth hormone that it started promoting the product even before the Food and Drug Administration had approved it. It stopped distributing videos to dairy farmers, dairy processors, politicians, scientists, and supermarkets only after the FDA ordered it to do so.

Records of the Formaldehyde Institute show that it spent hundreds of thousands of dollars a year on "communications" efforts in the early 1980s, largely on crisis management. At one meeting in 1981, members of the institute's communications committee patted themselves on the back for their success in persuading Donald Lambro, a syndicated newspaper columnist, to write a sympathetic column.[20] Lambro's column, which appeared in about 100 newspapers, lashed out at the Consumer Product Safety Commission for seeking to ban formaldehyde foam insulation. The Formaldehyde Institute also arranged tours, costing about $3,000 each, for its officers to meet with the editorial boards of newspapers in cities that they considered "hot spots."[21]

The Formaldehyde Institute invited Bob Reid, a senior producer for investigative reports at KNXT-TV in Los Angeles, and Mike Chaplin, the financial news editor of the *Los Angeles Times*, to speak at its annual meeting in 1981, in a panel moderated by Rich Good of Georgia-Pacific. "Mr. Reid touched upon a few pointers that would help Institute members deal with investigative reporters," the minutes of the meeting say. "It was similar to a media training session. He mentioned a few 'tricks' reporters may use to get the answers they are looking for."[22]

The worst crisis for the Formaldehyde Institute came in the early winter months of 1982, when ABC-TV's news program *20/20* started work on a story on the urea-formaldehyde foam insulation that had been blown into the walls of hundreds of thousands of homes during the energy crisis of the late 1970s. Homeowners in New England and the Great Lakes states now had asthma, dizziness, allergies, rashes, nausea, and memory loss. They had paid just a few hundred dollars apiece to have the insulation sprayed into walls; getting rid of it would cost tens of thousands of dollars.

Peter Lance, a correspondent for *20/20*, interviewed some of the victims. One of them, Michael Wagner of Bayville, New Jersey, told him of equilibrium problems, bad headaches, nosebleeds, strange psychological symptoms (such as an inability to handle stress), and, finally, a collapse and hurried trip to the emergency room. Lance's camera crew had videotaped workmen removing the hardened white chunks of formaldehyde foam—which doctors found to be the source of Wagner's illness—from his home at a cost of $20,000.[23]

Lance interviewed other families, real estate agents, officials of the Consumer Product Safety Commission, and Peter Breysse, a professor of engineering at the University of Washington who advocates sharply limiting formaldehyde levels indoors. He also sought to interview someone from the Formaldehyde Institute. Joel Bender, a DuPont scientist and lobbyist, agreed to appear, but only on one condition: that the Formaldehyde Institute be permitted to videotape the interview at the same time *20/20*'s cameras were running. "If *20/20* refuses us permission to simultaneously videotape the interview, it is likely that DuPont will pull Dr. Bender out," the minutes of a meeting of the Formaldehyde Institute show.[24]

DuPont got its wish, and 13 days after the airing of the *20/20* segment ("The Danger Within") on February 4, 1982, the formaldehyde manufacturers already had produced their own film. John Paluszek, then of Paluszek & Leslie Associates, one of the Formaldehyde Institute's two public relations firms, distributed a draft copy of "A Documentary on a Documentary" to the communications committee. Using outtakes of the two-hour interview that it had recorded simultaneously with *20/20*—at Bender's request—the counterdocumentary sought to show that ABC had slanted its story against formaldehyde.[25]

The institute had already sent a letter of protest to Roone Arledge, then the president of ABC News. It was also compiling a brochure "contrasting statements made in the telecast with other statements taped by *20/20*," which would be distributed to public health departments throughout the nation and to all members of the Formaldehyde Institute for distribution to their employees.[26]

Georgia-Pacific produced another film by splicing together pieces of the *20/20* segment with parts of the industry's simultaneously recorded interview and offered it to other formaldehyde manufacturers. The institute decided to review the interview again "for possible uses in the future, including at the annual meeting."[27]

In many respects, the corporate filmmaking festival failed to dull the impact of the *20/20* broadcast. A few days after the broadcast, the Consumer Product Safety Commission banned foam insulation—a plan that had already been in the works during the *20/20* shooting. Two congressional committees scheduled hearings on the problem. Two years later, the *National Law Journal* would identify formaldehyde lawsuits as the fastest-growing area of litigation in the country.[28]

Three months after its original program, *20/20* aired a short follow-up, showing a doctor, Gary Prisand, testifying before Congress about the ordeal of his wife. She had been stricken with rare nasal cancer—the same unusual type of cancer that consistently developed in rats exposed to formaldehyde in scientific tests.

"Recently I saw the television show *20/20* and listened in horror as I realized that our urea-formaldehyde foam insulation may have been the cause, not only of the tumor but also of the inability to heal and all the other post-operative complications," Prisand said. "With the knowledge we have now after hearing the *20/20* report and the literature we have collected, I must say I am shocked and appalled that it was the cause of the formaldehyde industry—not fate or an act of God—that caused this pain and heartache."[29]

Then Hugh Downs, the host of *20/20*, said: "Industry representatives who heard Dr. Prisand's testimony acknowledge the tragedy, but insist the connection between Mrs. Prisand's cancer and UF foam is coincidental."

It seemed like a public relations debacle, but the manufacturers were fighting a war of attrition at that point. Their losses were indeed massive; the foam insulation industry would die, even though the Formaldehyde Institute got a federal court to overturn the Consumer Product Safety Commission's ban of the product. Nonetheless, formaldehyde would continue to be used in a wide array of other home products, particularly particleboard, even though the *20/20* story included the problems of particleboard in mobile homes.

Over the years, manufacturers have learned that the in-your-face style of crisis management exemplified by the campaign against *20/20* is not always the best choice. Nowadays, they sometimes beat a strategic retreat in the face of bad news. That was the tack Jerry Levine of the Neighborhood Cleaners Association chose in November 1993 when, at an EPA public hearing on the dangers of perchloroethylene, a Manhattan woman told a horrendous story.

Chris Kruse explained that she had lived in her apartment building for 19 years and had smelled perc every working day since a dry cleaner had moved in downstairs five years earlier. One very cold day, when Kruse was eight months pregnant and was in her apartment with her two other children, the heating in the building stopped and the dry cleaner's pipes burst.

"It [the fumes] totally engulfed my apartment," Kruse said at the hearing. "It made me dizzy, more so than ever before. It was as though I were drunk with perc. I went downstairs and begged him to shut off the machine. And he was daffy, you know. I mean it was just literally pouring out the back of his machine. And he goes: 'I know, I know, it's killing me. I can't stand it. I'll turn it off in a half hour.'"

Kruse returned upstairs and opened the windows, but the fumes overwhelmed her. She and her children passed out, with all their windows open.

Levine and other industry representatives had aggressively cross-examined earlier witnesses at the hearing. But when he heard Kruse's story, Levine beat a hasty retreat to a fallback position that was easier to defend.

The dry cleaner in Kruse's building "operated in a way which was totally irresponsible, [and] inconsiderate because he operated with a machine that was so seriously malfunctioning it could have fumigated a whole block of houses, not just one apartment," Levine said. "He was operating on a machine that was made to condense perc that wasn't condensing any. So, there was no place for it to go but everywhere. But that isn't the rule. That's the exception and very unusual. That dry cleaner is one of the type that we probably would be better off without in this industry. . . . If he got regulated out of business I don't think any of the dry-cleaning industry people in this room would care one little bit, because he is the one who makes it bad for all of us."[30]

Putting It All Together

Over the years, the public docket room at the EPA's headquarters in Washington has been the repository of millions of letters. But the clerks who work in the crowded room had never seen anything like the deluge that began in 1994 in response to the agency's special review of atrazine. There were letters from politicians, from researchers, from commodity groups, and from cooperative extension agents. And, overwhelmingly, there were letters from farmers.

The letters were still trickling in during the summer of 1995, even though the public comment period had ended that March. The total, as of July 6, 1995, was staggering: 87,721 comments, all but 2,293 of them from farmers and other private citizens. "It's by far the most we've ever received" on a pesticide issue, says Joe Bailey of EPA, the manager of the atrazine special review.

By all indications, the deluge was an overwhelming mandate for the continued, unrestricted use of American agriculture's top-selling pesticide. It was also an awesome display of the chemical industry's financial muscle and public relations savvy. The paper avalanche—letters about atrazine filled 49 accordion-style files in the EPA docket room—was the result of an elaborate public relations strategy hatched by Ciba-Geigy and one its PR firms, Ceres Communications of Burnsville, Minnesota, a subsidiary of Shandwick Public Affairs, a multinational PR firm.

Ciba-Geigy got plenty of help from the farm press, from rural political leaders who had benefited from the company's charitable work, and, above all, from farmers. But the company did not stop there. It got university research departments, trade groups, cooperative extension offices, and other organizations to defend atrazine. Ciba-Geigy also blanketed agricultural magazines and broadcasts with ads urging farmers to speak up for atrazine, passed out postcards and form letters at community meetings, sent letters to tens of thousands of farmers, and made sure that its sales force personally contacted many farmers. Because many farmers use personal computers to obtain the latest farming news, Ciba-Geigy even worked out a way that farmers could, at the touch of a button, print out a form letter opposing atrazine restrictions that they could then send to the EPA.

Ciba-Geigy will not say how much it spent on the letter-generating campaign. "In relation to the amount of money we spend advertising other products, it was not substantial," says Robert Clark, a spokesman for the company. "We did things like direct mail, press releases. Growers don't readily find out about a special review. There's not a lot of time to comment, and we felt we needed to let them know what was happening—that if you use the product, and you're concerned, then let EPA know about it."

Ciba-Geigy was spending heavily for advertising and public relations even before the atrazine campaign began. In 1993, for instance, the company was the only agribusiness client of Martin/Williams Advertising, Inc., which reported agribusiness revenues of $1.15 million and told *Agri Marketing* magazine that the ads were for Ciba-Geigy's atrazine and metolachlor herbicides.[31] That same year, fees from Ciba-Geigy were the "significant majority" of the $866,000 in agribusiness revenues earned by the PR firm Ceres Communications, according to Bob Rumpza, Ceres's managing director. By 1995, the atrazine letter-generating campaign was under way, and Ceres's agribusiness revenues had hit $1.031 million—most of it from Ciba-Geigy.[32]

"They often look to us to work in the area of working with the agricultural media to spread the word about any given subject," Rumpza says. "The client decides what needs to be said and we help them say it." The atrazine campaign was important, he adds, because "people needed education. They needed to know the prescribed means of submitting something to the [EPA] docket—the address and the docket number."

Ceres and Ciba-Geigy also tried to make sure that the letters were phrased for maximum impact. The company's advertisements and pamphlets included a list of "key facts" that "may help you in making your points to the EPA." Without any apparent sense of irony, the Ciba-Geigy literature followed the detailed list with the tag line: "It's your EPA. Tell them what you think!" A more accurate slogan, perhaps, would have been, "Tell them what Ciba-Geigy thinks."

The vast majority of the letters were form letters, in varying stages of disguise. Many farmers simply mailed the EPA the letters they had received from Ciba-Geigy, scrawling their signatures in the margins. Joe Godwin of Screvin, Georgia, scribbled "Leave our way of life alone" atop a letter he had received from a Ciba-Geigy executive and sent it to EPA.[33] Others sent back the same typewritten form letters or copied Ciba-Geigy's suggested language by hand. Even the personalized letters frequently betrayed the fingerprints of a public relations campaign: a preprinted envelope, perhaps, or an out-of-state postmark.

"With the petitions and postcards you can clearly tell what's happening—you get letters that say the exact same thing," the EPA's Bailey says. "We don't really put a lot of effort into looking at people's comments who say: 'We use the stuff. It works really well. Please don't cancel it.' Those don't lend any additional evidence either for or against its risk."

That does not mean Ciba-Geigy wasted its money. Logging in the letters can take the EPA months, further stretching out a special review process that already takes several years to complete. "People have their Xerox machines running," the EPA's Joseph Reinert says. "They have their chain letters out to their growers. It slows the process down because you have to sort through it all. We got enough comments on atrazine to fill a couple of football stadiums."

More significant, the deluge is a tangible sign of the political muscle of the pesticide industry and a warning to EPA of the heat it would have to take if it cracked down on atrazine. Campaigns like Ciba-Geigy's inevitably present a skewed assessment of public opinion that, Bailey's protests to the contrary, cannot help but influence the EPA. Organizations that favor tougher restrictions on atrazine cannot afford major advertising campaigns. "Ciba's doing a real heavy campaign and it gets me mad," says J. Alan Roberson of the American Water Works Association. "How am I going to counter that? I can get maybe 12 or 15 guys to write letters."

Just as Ciba-Geigy did, the manufacturers of perchloroethylene used all of their public-relations muscle to wage a campaign that most Americans are not even aware of—even though evidence of the industry's success is as close as the nearest closet or wardrobe.

The typical American closet contains scores of miniature advertisements for perchloroethylene. Thanks to one of the most effective campaigns the industry has waged to advocate greater use of the chemical, federal law permits—and even encourages—DRY CLEAN labels in dresses, suits, sweaters, and other garments that could be safely cleaned in water. Regulators and environmentalists believe that these tiny warnings stand as a significant obstacle to professional cleaners turning away from chemicals and to carefully controlled water cleaning, a computer-aided technology that is being used in more than a hundred shops across the United States.

"Dry cleaners fear massive liability for cleaning clothes with water when care rules specifically say 'Dry Clean Only,'" Bonnie Rice, a Greenpeace activist, said in a letter to the Federal Trade Commission. "They are extremely reluctant to use anything other than perc-based methods on such garments, even when they know the garment can be safely wet-cleaned."[34]

Under federal regulations that have been in effect since 1981, clothing must have "either a washing or dry-cleaning" instruction and "need have only one" if either method of cleaning will do.[35] This edict emerged from the Federal Trade Commission after years of fierce and combative lobbying over clothes-labeling requirements by the sometime-competing interests of the clothing and textile makers, retail stores, home laundry soap producers, and, of course, the dry-cleaning industry.

Ordinary consumers spoke, too, most notably in waves of letters that would arrive in Washington after an article in a national magazine or local newspaper. But it is nearly impossible to sort out which letters were genuine expressions of concern from grass-roots America and which emerged from a sophisticated big-business technique that has become known as "astroturf" lobbying. In 1976, for example, reams upon reams of petitions signed by dry-cleaning customers (but distributed and gathered by the Neighborhood Cleaners Association) were delivered to the FTC, calling for the most stringent care-labeling scheme possible. The agency estimated that it received petitions with more than 45,000 signatures and thousands more letters and cards.[36] They arrived day after day, like the bags of mail dropped on the judge's desk in *Miracle on 34th Street* to convince him that Santa Claus was real.

The public's desire for clothing-care labeling was not quite as mythical as Saint Nick. But the flood of industry-sponsored lobbying has prevented government regulators from addressing some of the biggest concerns of the average consumer. As early as 1971, when the FTC was considering its first formal regulations on washing labels, it noted that some garments with DRY CLEAN ONLY labels could be washed at home at lower cost, a situation it called "low labeling." The agency attributed the practice to the textile industry's wish to avoid the cost of testing whether clothes are washable in water.[37]

Despite repeated attempts by the FTC's staff since the 1970s, the agency could not finalize a regulation that would tell consumers when a garment labeled DRY CLEAN could also be washed. Every time such a proposal would surface, the International Fabricare Institute, the Neighborhood Cleaners Association, and other dry-cleaning interests would weigh in, bringing their store of petitions from customers. They never said that they opposed the "alternative" care idea; in fact, they embraced it even more wholeheartedly than the agency could have imagined. But the only fair way, they said, was to put WASH OR DRY CLEAN on *all* clothing, towels, home items—even those that normally would be machine-washed. After all, most washable garments also can be dry cleaned.[38]

8

Assessing the Alternatives

The old Monsanto slogan, "Without chemicals, life itself would be impossible," neatly sums up the chemical industry's strategy of making its products unavoidable. No matter whether the intended audience is a federal bureaucrat, a politician, an investor, a juror, or someone on the other side of a television screen, the chemical industry's public relations machine invariably pounds home the message that there are no reasonable alternatives to toxic products, that all other options are costly, impractical, or ineffective.

For most synthetic chemicals, the evidence of at least some risk is irrefutable. Consequently, a toxic product's survival in the marketplace depends on making the case that, while it may pose a health risk, the risk is worth taking because it is so small, or that there is no better or cheaper way to weed a cornfield, clean a dress or suit, or make a kitchen cabinet. Our food supply, our desire for an environment clean of germs and biological contaminants, our need for affordable housing all demand that we use ever-increasing amounts of toxic chemicals—or so the industry tells us in a relentless barrage of advertising, lobbying, and public relations.

That argument withers under scrutiny, however. In the same way that test-tube wizardry helped usher in the chemical revolution a half-century ago, high technology is showing us that there are indeed alternatives to chemical dependency. Dry cleaning could be abandoned in favor of professional water-based cleaning with computerized control of temperature, agitation, and drying. Chemical-based agriculture could give way to organic practices in which farmers use computer databases and high-tech monitoring to help them protect crops without weed-killers and insecticides. At least one company has already figured out how to make a particleboard without formaldehyde; more are sure to follow.

There is no doubt that there is pent-up consumer demand for safer products. Even industry trade publications have acknowledged it. "Surprisingly, however, one of the strongest voices calling for alternative compounds—the fastest-growing demand for greener chemicals—is coming not from environmental groups or government regulators but from the customers," an August 1995 article in *Chemical Week* observed.[1]

That demand is a threat to the chemical industry, and manufacturers are fighting back. In October 1995, for instance, the Chemical Specialties Manufacturing Association declared war on baking soda and vinegar. The association, which represents the makers of most cleansers, pesticides, and other chemicals used in homes, launched a public relations crusade in which it urged consumers to reject common, benign, inexpensive alternatives such as baking soda and vinegar and instead return to the supermarket aisles where members of the association market their "smart choices for cleaner living."[2]

When combined, baking soda and vinegar will bubble but will not clean a clogged drain like a chemical drain cleaner, one of the association's pamphlets says. The chemical industry's products are "carefully tested, evaluated, and optimized for performance, effectiveness, stability, human safety, and minimal environmental impact" and are labeled to explain ingredients, product use, storage, and disposal directions. "Home mixtures don't offer any of this vital information."

There is something else home mixtures do not offer: profit margins that can compete with those of synthesized chemicals protected with patents. So there was no countercampaign to answer the association's assertions. There was no organized effort to point out that, yes, chemical drain cleaners work better on clogs, but if consumers regularly washed out drains with the baking soda–vinegar combination they could prevent clogs in the first place. No one distributed pamphlets noting that the safety tests the industry touts are conducted by and for manufacturers and that product labels rarely tell the full story. Indeed, a consumer would be lucky to find the word "formaldehyde" on even half of the products in his or her home that actually contain it, because the chemical is usually added to the mix as a minor ingredient that by law does not have to be labeled.

Considering the chemical industry's dominance of research, regulation, and information, it's no wonder that most users of alachlor, atrazine, formaldehyde, and perchloroethylene, for example, have never heard of practical and economical alternatives that do not depend on the heavy use of synthetic chemicals, and in many cases do not require them at all. As a result, people are not asking for chemical-free wood products at the local lumber yards, most farmers have not adopted new tilling methods, and there are probably fewer than 200 professional wet cleaners scattered around the country.

But now the good news: Some builders, farmers, and cleaners have discovered that life *is* possible—even better—without some toxic chemicals.

Formaldehyde Alternatives: Healthy Homes

By the time John and Lynn Marie Bower finished remodeling the master bedroom of their 1850 farmhouse outside Lafayette, Indiana, in the late 1970s, "I was finished, too," Lynn recalled.

Her hair was falling out, she had developed acne and other skin ailments, she could not keep food down, she had dizziness and hallucinations, and, for several years, she was bed-bound. Worst of all, she had developed a sensitivity to virtually all chemicals, experiencing allergic reactions to common perfumes, shampoos, and soaps.

The Bowers could not pinpoint exactly what caused her problems; it was somewhere in the soup of chemicals they had unwittingly created in their home by installing new flooring, walls, furniture, and carpeting. Lynn suspected that formaldehyde—with its tendency for making people hypersensitive to all chemicals—was one of her bigger exposures. She knows now what she did not know then: Formaldehyde could have been in any of the adhesives, caulks, particleboard, plywood, or paneling they had brought into their home—even any permanent-press fabrics.

John Bower, then a designer for an engineering firm, changed how he approached his work because of his wife's illness. Out of necessity, he began seeking out building materials and furnishings that were free of toxic chemicals, so that they could move into a new home where Lynn could breathe again. In the process, he discovered that, while affordable alternatives are available, it is difficult and time-consuming for the average consumer to find them.

Most experts on nontoxic building materials say, for example, that you can forget about going to the local hardware store or lumber yard for help. Such stores provide little information on product content and alternatives, and most products that are truly low in toxics are not produced by companies that have big distribution networks that ensure their presence on the shelves. Even products one might think are nontoxic—solid wood, for example—are often treated with such chemicals as preservatives, insecticides, or fungicides.

Not only are the alternatives difficult to find, but labels like "green" and "environmentally friendly" just add to the confusion. True, the Federal Trade Commission is supposed to ensure that such claims are true. However, there is little that the agency can do about claims that are true but misleading. For example, most paint manufacturers, led by Glidden, a division of the British chemical giant ICI, have developed paints that are low in chemicals and odors called "zero-VOC" paints. A VOC is a "volatile organic compound"—in other words, a compound that the EPA regulates because it contributes to smog, one of the earliest concerns of environmentalists in the 1970s. But most people do not worry too much about the creation of smog inside their homes; they worry about toxics. And most "low-VOC," or low-smog-producing paints, contain toxins such as crystalline silica and acetone, both suspected carcinogens, and ammonia, a harsh solvent.

This galls Benjamin Goldberg, the president of American Formulating and Manufacturing of San Diego, which has been producing paints free of these and other toxins for 15 years. Goldberg's company was started by chemists who left the chemical industry because they believed it was possible to reduce the toxins and irritants in paints and other consumer products. But American Formulating and Manufacturing cannot market its paints as zero-VOC because they contain such chemicals as propylene glycol, a vegetable-derived substance that is not toxic yet is a precursor to smog, and thus a VOC. A consumer trying to avoid toxics would be better off buying this low-VOC paint instead of another company's zero-VOC paint, Goldberg argues. Although paint companies are using the zero-VOC label as a marketing tool, Goldberg says that "it is meaningless if you want information about toxicity."

John Bower knew how to look for products because he was an

engineer. After amassing the information on household cleaners, cosmetics, wood products, upholstery, carpeting, and ventilation systems that they needed to move on with their own lives, the Bowers began making their work available to others. They wrote books on chemical-free homebuilding and nontoxic household products, and John began designing and building chemical-free homes for people who suffer from the same sensitivity problems that had incapacitated Lynn.

Formaldehyde can be avoided, the Bowers say, but it takes some effort. Almost all of the particleboard, plywood, and medium-density fiberboard on the market today contains one of two types of formaldehyde glue. Most composition-wood products for use inside the home are held together with urea-formaldehyde glue. When mixed with formaldehyde, urea—a synthetic version of the acid in urine—creates a seal so stiff that it gives what nature rarely provides: a wood that does not warp. Unfortunately, urea and formaldehyde also have something of an open relationship. Formaldehyde can be coaxed out of its urea bond by humidity's slightest touch. Wood products for decks, siding, and other areas outside the home are usually glued together by the stronger bond that formaldehyde creates with phenol, a petroleum product. Phenol allows less gas to escape into the air, although the footloose formaldehyde can never be reined in completely. Why do more wood-products manufacturers not use phenol? As a petroleum product, phenol is much more expensive and tends to darken wood.

In his book, *The Healthy House*, John Bower argues that products containing urea-formaldehyde "should never be considered for use in an ecologically safe house, and those containing phenol-formaldehyde should be avoided if at all possible." Although the wood-products industry has changed its formulas since the 1980s to use less formaldehyde, Bower writes, "the amount of formaldehyde released is still too high for this material to be considered in an environmentally safe house."[1]

The National Particleboard Association disagrees. It points to a study it jointly conducted with the federal government of a single test home heavily furnished with its products.[2] "The results support what industry has been saying for years," Richard Margosian, the association's president, said. "Formaldehyde emission levels from composite-wood products are a fraction of what they were 20 years ago, and when these products are used in housing, the resulting formaldehyde levels are extremely low."[3]

Whom should the consumer believe? The truth, if one pieces together facts that both sides accept, is that most people cannot detect formaldehyde in a home if the flooring and furnishings conform to the industry standard, if the ventilation is good, if the temperature and humidity are low, and if drywall and other porous materials do their usual job of absorbing formaldehyde gas. Some people still will be able to smell formaldehyde gas, and will be irritated by it, and it still will be there, released from the wood and the walls and the fabrics in the home over weeks, months, years—whenever the temperature and humidity rises.

Shouldn't consumers be able to choose not to breathe this known irritant and probable carcinogen—at whatever level? The work of the Bowers and others is aimed at providing that choice.

Solid wood, which costs more than wood-veneer products, may be an alternative in some places in the home. Solid poplar can even be less expensive than the top-of-the-line walnut-veneer products on the market, Bower says. But he points out that much solid wood today is chemically treated to make it more resistant to fungi, insects, and fire. Redwood and cedar, however, have natural decay and termite resistance and could be chemical-free alternatives.

It is difficult to find formaldehyde-free particleboard. Masonite Corporation of Chicago offers a product that is bonded by heat and pressure so that the natural lignin in the wood holds the wood particles together. Sometimes, however, formaldehyde and other chemicals are added to Masonite to improve its strength and moisture resistance. (Masonite Corporation was a member of the Formaldehyde Institute.) At least one U.S. manufacturer, Medite Corporation of Medford, Oregon, produces formaldehyde-free particleboard. Hank Snow, a vice president of the company, says that the urea-formaldehyde resin is replaced by a "poly-urea resin matrix." Because so few chemical companies make the alternative resin (at one point, Medite had to import it from England), Medite's particleboard products, Medex and the less expensive Medite II, cost about 30 percent more than the equivalent formaldehyde wood products, which Medite also sells. Despite the cost, architects, hospitals, and libraries often request the formaldehyde-free product, Snow says. *Environmental Building News*, a newsletter based in Brattleboro, Vermont, reports that Medite II also is widely used in art galleries and museums whose curators are concerned that formaldehyde gas damages exhibits.[4]

Some homeowners and remodelers settle for reducing formaldehyde gas instead of eliminating it. Sergio Zori, an architect in Sea Cliff, New York, began specializing in environmentally conscious building design because of his own allergies and asthma. Zori says that if he has to use a formaldehyde product, he uses the type that out-gasses the least: phenol-formaldehyde. To further reduce fumes, he coats it with a sealant designed to cut down on gas. One brand is marketed by American Formulating and Manufacturing and another by Palmer Industries, Inc., of Frederick, Maryland.

Nadav Malin, the managing editor of *Environmental Building News*, says that some people might be sensitive to sealants as well. He recommends that anyone who is thinking of building or remodeling a home take samples of products they intend to use, place them in a jar, and test how they react to breathing them.

Testing a home for gases like formaldehyde is difficult; once again, do not expect to find a tester in the local hardware store. The best bet is to look in the yellow pages under industrial health and safety supply companies; they usually carry relatively inexpensive monitors (about $50 apiece) that can detect low formaldehyde levels. Similar monitors are available for perchloroethylene, the dry-cleaning chemical, and a few other common chemicals. Many people with indoor-air-quality problems choose to call on their state health departments, which often have testing equipment and personnel.

John Bower says in his book *The Healthy House* that he believes that for a house to be truly free of toxic chemicals, one must completely eliminate what he calls the Big Three: composite and treated woods, carpeting, and combustion appliances (such as wood stoves, oil furnaces, and other heaters, which can generate deadly carbon monoxide). Carpeting not only holds dust, mold, and pesticides, but the finish, adhesive backing, and batting on most brands contain hundreds of chemicals that leach into the air.

Insulation also can be a troublesome source of toxic substances in the home, even though urea-formaldehyde foam insulation has virtually disappeared from the market, a tribute to the power of consumers. Once Americans heard about the Consumer Product Safety Commission's decision to ban urea-formaldehyde foam, the market for it all but dried up—even after the Formaldehyde Institute succeeded in getting a court to lift the ban.

But any consumer shopping for insulation soon learns that it is difficult to find hazard-free products. One of the most popular insulating materials, fiberglass, often includes formaldehyde as a binder. Furthermore, fiberglass itself has been shown to be carcinogenic. Some builders believe that the only nontoxic insulation alternatives are straw, feathers, and sawdust; all, however, are prone to insect attack. Natural cork is another alternative, but it is relatively scarce and expensive. What's more, some products contain "natural" cork scraps that have been agglomerated (bound together) with—you guessed it—formaldehyde glue.

Some environmental architects are beginning to recommend Air Krete, a foam-insulation product manufactured by Palmer Industries. Instead of using formaldehyde as the binder, Palmer uses salt and minerals extracted from sea water. Zori, who has insulated his own home with Air Krete, estimates that it costs twice as much as fiberglass insulation. He still recommends it, however, and believes it is more effective and long-lasting than fiberglass insulation.

Price, of course, is always an important issue. Toxic-free building materials generally are more expensive, not only because of the ingredients but also because the companies producing them are so small that they do not realize economies of scale and because there are few distribution networks for such products. Still, Goldberg, of American Formulating and Manufacturing, says that his company's low-toxicity products are competitively priced with top-of-the-line mainstream products.

From these few examples, it is obvious that it is difficult for consumers to make nontoxic choices in building products. Thad Godish, the director of the Indoor Air Research Laboratory at Ball State University, says he has found it virtually impossible to figure out which kitchen cabinets have finishes that emit formaldehyde and which do not. And Sara Schotland, a longtime lawyer for the formaldehyde industry, warns that consumers might turn to alternatives that are as hazardous as or more hazardous than formaldehyde. Some types of particleboard, for example, are made with isocyanate resin, whose hazards are well known and great. She also points out that particleboard creates a use—a recycling—of scrap wood for an industry that has been under fire for overcutting the nation's forests.

These are difficult and important issues. But toxic-free building products are available, though the market in them has not yet had the chance to grow, mature, and flourish. The question is: What would have happened if all the engineering ingenuity, money, and political muscle that at least 40 corporations invested over 20 years in protecting formaldehyde's place in the market had been devoted to driving it out?

Herbicide Alternatives: Healthy Farms

Like thousands of other Iowa farm kids, Vic Madsen grew up with a clear vision of what a cornfield ought to look like—as clean and monochromatic as a newly vacuumed carpet. When Madsen, now 49, first started farming his own 400-acre spread near Audubon, Iowa, he adopted conventional farming techniques without a moment's pause. He sprayed his fields over and over with high doses of herbicide mixtures, including ones that contained atrazine and alachlor, and he applied insecticides even if there were no signs of insect damage. Heavy herbicide use, he explains, is deeply rooted in the culture of Iowa. Ego, peer pressure, and short-term financial pressures are all powerful forces that drive Iowa farmers to make their fields "clean," or weed-free. "Part of your identity is tied up in that land," Madsen says.

But behind those powerful forces is the hidden hand of the chemical industry. The aggressive marketing of chemical manufacturers, Madsen says, has warped the culture and the economics of farming. "The Dows, the Geigys, the Monsantos, and the [American] Cyanamids have excellent marketing firms," he says. "They conduct numerous studies and they probably know how many times their average customer drinks coffee a day, how many times he eats, what color shirts he likes to wear."

After years of heavy spraying, Madsen says, he eventually came to realize that there is another way to farm—a way that saves money, protects the environment, and preserves human lives. He started cutting back on herbicides and synthetic fertilizers and relying more on innovative tilling techniques and mechanical cultivation. He also discovered Practical Farmers of Iowa, a 500-member organization united by a common drive to defy convention and try innovative ways of farming that are both profitable and environmentally sound.

Unlike some other members of Practical Farmers of Iowa, Madsen has not eliminated chemicals from his farm. But he has cut his use of synthetic fertilizers in half, and of herbicides by a third, while increasing his per-acre yields and profit margins. And he is planning even deeper reductions in his use of chemicals.

In fact, a respected series of studies by the organization suggests that

Madsen and other farmers ought to consider going cold turkey on herbicides. Alachlor and atrazine are among the cheapest of all pesticides, yet the studies suggest that they are not worth the cost. Since 1987, members of Practical Farmers of Iowa have collaborated with the Iowa State University Cooperative Extension in studies that have measured corn and soybean yields in fields where herbicides are used and in fields where a mechanical alternative known as ridge tilling is used. With an elaborate formula that takes into account the extra labor costs of mechanical cultivation as well as the cost of chemicals, the studies have estimated the cost per acre of each type of farming.

Through 1994, Practical Farmers of Iowa had conducted 21 corn trials and 30 soybean trials, each involving six fields with similar characteristics (to guard against statistical anomalies). So many fields have now been tested that the studies are widely cited in scientific agricultural journals because of their statistical reliability. The studies show that yields, on average, are virtually the same: for corn, 129.9 bushels per acre in organic fields and 129.5 bushels per acre in herbicide fields; for soybeans, 42.4 bushels per acre in organic fields and 43.2 bushels per acre in herbicide fields. For both crops, herbicide-free farming offers major financial savings: $5.82 per acre for corn and $5.95 per soybeans.[5] That does not include, of course, the long-range costs of pesticide use: environmental damage, increased weed and pest resistance, and the destruction of such agriculturally beneficial insects as honeybees.

"What we've shown is that it can work, and that on a per-acre basis it can be very competitive economically," says Rick Exner of the Iowa State University Cooperative Extension. "The question is, why isn't everybody doing it?"

Researchers at the Rodale Institute in Emmaus, Pennsylvania, ask the same question. In 1981, scientists there launched a long-term study aimed at determining how difficult it is for farmers to go organic. To compare the two methods, researchers chose three fields that for years had been continuously farmed using pesticides, synthetic fertilizers, and the other tools of conventional farming. In one field, they made no changes in farming practices. In the other two, they went cold turkey on chemical insecticides, herbicides, and fertilizers. To fertilize the organic fields naturally, they instituted a crop rotation system and planted clover and other cover crops during off-seasons. They also added manure to one of the organic fields.

For the first four years of the experiment, the researchers found, chemical farming had the upper hand. Corn yields were, on average, slightly more than 20 percent higher on the conventionally farmed fields than on the organic fields from 1981 to 1984. But the gap closed quickly as natural nitrogen built up in the organic fields, and from 1985 to 1990 yields were virtually identical—in fact, the organic fields that got manure actually produced slightly more corn than the conventionally farmed fields.[6]

Rhonda Janke, the former research director of the Rodale Institute and now an associate professor of agronomy at Kansas State University, says that farmers can readily avoid a rough transition by easing out of chemical farming, instead of going cold turkey, and by choosing smart ways of rotating crops and tilling soil. "We've shown that you can get the same yield and you can do it for the same money without herbicides, but your cropping system's going to look different, and the way you manage it is going to look different," she says.

The chemical industry dismisses these small-scale studies as wildly unrealistic. The Practical Farmers of Iowa study "doesn't pass the laugh test, because farmers are some of the most sophisticated businessmen that we have in the country," Darrell Sumner, Ciba-Geigy's manager of health and safety issues, says. "If they were able to save $20 an acre or $10 an acre [by dropping herbicides], they would." To support their case, pesticide manufacturers point to studies such as one conducted by agronomists at Texas A&M University, who concluded that corn yields would fall 53 percent if farmers stopped using synthetic fertilizers, weed-killers, insecticides, and fungicides.[7] But that study, like most of the others cited by the chemical industry, assumes that farmers would simply abandon chemicals without adopting innovative ways of rotating crops and tilling soil. "An atrazine study conducted by a chemical company would say: 'Here's corn. What happens when we take the atrazine out?'" Janke says. "And of course, it [the yield] falls flat on its face."

Conventional corn farmers almost always spray herbicides early in the spring before the corn stalks have emerged from the soil. Then they do not have to worry about attacking weed problems later in the growing season. On the other hand, organic farmers have to be ready to return to their fields to get rid of weeds with mechanical cultivators. If the fields are too muddy to plow and they are forced to wait, the weeds could grow so high that cultivation is impossible, driving down crop yields. "In some years, in good years, that sort of [low-]input farming can produce about as much profit as conventional farming, but if the weather doesn't agree, then it can go south very fast," Sumner says. "So there will be years where the yield doesn't occur, and then it becomes very unprofitable and infeasible."

Farmers often are simply too busy during the growing season to put in the extra time needed to remove weeds mechanically, says Leonard Gianessi, a senior research associate with the National Center for Food and Agricultural Policy, a Washington-based research organization that was started by agribusiness interests, which still provide about half of its funds. Just down the road from the Rodale experimental farm in Pennsylvania, he says, is a larger operation run by an Amish farmer who always uses atrazine. "The average-sized corn and soybean farm is 800 acres," Gianessi says. An organic farmer "would be on the tractor for two weeks just to make one damn cultivation pass."

Weaning farmers off herbicides, critics argue, poses other problems as well. If a farmer decides not to spray weed-killers before crops emerge but then finds that the fields are too muddy to plow, the only alternative may be aerial spraying—which on windy days can drastically increase a farmer's exposure to chemicals, Gianessi says. And while organic farming might work well in Iowa and Illinois, where soils are rich and farmers can afford to put in the time needed to learn new techniques, it is less likely to succeed in places where growing conditions are not ideal. Some studies of alternative farming techniques "probably do show that with little or no [chemical] inputs you can have equivalent yield, or maybe not too big of a yield loss, and that's fine," says Andrew Klein, a research manager at Monsanto. "But if we're going to feed the world, you need to have broadly applicable technologies and you have to have better yields, not just the same yields." Declining crop yields, industry scientists argue, could also speed the destruction of rainforests and other environmentally sensitive lands by causing them to be plowed under for food crops.

Some argue that frequent plowing—the only way to remove weeds without chemicals—inevitably increases soil erosion. Monsanto, Ciba-Geigy, and other herbicide manufacturers, in fact, have eagerly embraced "no-till" farming techniques, in which farmers abandon all plowing—and rely on herbicides—in the hope of conserving topsoil. Both companies routinely sponsor seminars that teach farmers no-till methods, pitching their weed-killers in the process. The need to combat soil erosion not only has spurred weed-killer sales but also has given manufacturers an opening to co-opt the concept of "sustainability," a term that environmentalists use to signify actions that will help sustain people and the environment in the long term. "We really can't stand the same amount of soil erosion, the loss of our topsoil, in the next 100 years that we had in the last 100 years," Sumner says. "Herbicides are necessary for no-till, to be able to have sustainable agriculture."

But many people do not think so, including Dennis Keeney, the director of Iowa State University's Leopold Center for Sustainable Agriculture. Over time, he says, weeds inevitably develop genetic resistance to weed-killers, just as insects eventually develop resistance to insecticides. "A tremendous resistance level is being built up by weeds," Keeney says. "Glyphosate [another Monsanto herbicide] is the only one that has been able to hold its own. It [switching from one farm chemical to another] is a short-run strategy. In the long term, we're just going to keep having problems. Nature has really got it figured out and is thoroughly enjoying the battle. We just can't stay ahead."

Keeney and others acknowledge that operators of huge farms may find it difficult to make time for the labor-intensive job of mechanical cultivation and that organic farmers must act quickly to respond to changing conditions. "It's a system that requires a little more timeliness, and a little more time—maybe another ten minutes per acre—and that goes against the pressure people are feeling to farm more acres less intensively," says Exner of the Iowa State University Cooperative Extension. "The pressure is to limit the amount of management you have to put in per acre in order to be able to handle more acres."

Critics of chemical-dependent farming suggest that the concentration of more and more farmland in the hands of fewer and fewer farmers has not been good for rural America. "The question is, are we going to have a very few large farms controlling things or are we going to retain family farms?" says Chuck Hassebrook of the Center for Rural Affairs in Walt Hill, Nebraska, a research and advocacy group for small farmers. "The more we move agriculture toward total dependence on pesticides to control weeds, the more it facilitates the concentration of wealth into a few very large operations. When all you have to do is spray on some product, one person can cover a whole lot of acres. You use that product to replace the management skills of the farmer in weed control.

Less of the money goes to corn farmers for their time, and more of the
money goes to the Monsantos of the world for their product. More of the
dollars are captured by the chemical industry, and less for the family
farm sector."

Hassebrook and others endorse an alternative model of agriculture
that is well suited to low- or no-chemical farming in which more farmers
work on smaller plots of land, where they plant different crops from year
to year as market conditions dictate, instead of always growing corn,
soybeans, or some other staple. In a way, the system is a throwback to
the old days, when farmers used manure, crop rotation, and off-season
cover crops to enrich their soil and hold it in place and relied on
mechanical cultivators to dig weeds. Modern organic farmers do all those
things, too, but there are crucial differences on the organic farms of
today. Technology gives growers the information they need about how to
manage crops efficiently and keep yields high, and modern cultivating
machines are much faster and more effective than their progenitors.

"There's this image that it's horse-and-buggy agriculture, and in the
overwhelming number of cases that is not true," says Garth Youngberg, a
pioneer of modern organic farming and the director of the Wallace
Institute for Alternative Agriculture in Washington. "It's almost impossible to get people to understand what's being accomplished on some of
these organic sustainable farms. It's shifted away from being a hippie
thing, but there are a lot of people out there who haven't shifted their
view. There's a story about a conventional farmer who toured a well-
managed sustainable farm, and as he was getting on the bus he said, 'I've
seen it, and I still don't believe it.' It's a psychological barrier."

Another myth is that alternative farming inevitably increases soil
erosion. Modern organic farmers can keep erosion to a minimum by
planting off-season cover crops, rotating cash crops, and using innovative
cultivation techniques such as ridge tillage, Janke says. Abandoning
plowing might make sense for farmers who are vulnerable to erosion
because they plant the same crop year after year. But very few
conventional farmers who have adopted the no-till methods championed
by pesticide makers have actually stopped plowing entirely. "I would
challenge you to find a single farmer who is doing continuous corn with
no-till," Janke says.

As Vic Madsen and many others have shown, farmers can step off
the pesticide treadmill and still turn a profit. Of course, with pesticide

manufacturers so deeply enmeshed in the fabric of rural community life, it takes some effort. "It takes planning, and a real understanding of the process," Janke says. "But it works."

Solvent Alternatives: Healthy Cleaning

It is not primarily for environmental or health reasons that Richard Simon believes in cleaning fine clothes with water, as he, his father, and his grandfather had been doing in his native England for 75 years. "You have to ask yourself, if you had the choice, would you dry clean the clothes you wear closest to your body?" Simon says. "And if the answer is yes, won't you please sit on a different carriage, because I really don't want to be next to you."

The fact is that although perchloroethylene does a fine job of dissolving grease (it originally was developed as a metals degreaser), it does not work so well on sweat, blood, or any of the organic soils that come from the human body. The dry-cleaning industry acknowledges this, which is why they rev up, or "charge," their machines by adding a small percentage of water to the dry-cleaning mix. Some people might notice this in the tugs or buckles on their wool suits, if they have been dry cleaned frequently enough in water-charged machines. Water makes the lining shrink faster than the jacket material.

Simon argues that water cleaning is better and fresher for clothes than chemical cleaning. But the temperature, agitation, amount of water, and speed of drying have to be carefully regulated for each different type of fabric. Simon's father and grandfather would put their hand in the cleaning drum every five minutes or so to check if the water was right for the fabric, like a parent testing the temperature of a baby bottle. Dry-cleaning machines eliminate the bother and expense of human judgment, because they rely on chemicals that neither shrink nor expand fabric.

Just since 1990, however, advances in computer technology have paved the way for water-cleaning systems that do not force professional cleaners to be nursemaids to their machines. Aqua Clean Systems, Inc., of Inwood, New York, has a high-tech, computer-driven system with a variety of settings for different types of fabrics. The company has sold its system to about 700 cleaners over the world, including about 200 in the United States. Neal Milch, a vice president of the company, which is a subsidiary of the Swedish manufacturer Electrolux A.B., known best for its vacuum cleaners, says that it was government regulation that forced the development of the wet-cleaning machines—not U.S. regulation, but regulation in Germany, which in recent years has ordered dry cleaners to curb their emissions to protect workers from perc fumes.

The German government requires perc fumes to be so low—one-tenth of the current permissible exposure level for workers in the United States—that the industry needed to look for alternatives to perc. And Electrolux, which already made commercial washing machines for professional shirt and sheet laundering, developed a technology to clean wools, silks, and other fine garments with water.

"It is nothing more and nothing less than automated hand washing," Milch says. "You could hand-wash a wool suit at home. The problem is, it would take you nine hours to finish it [to block and dry it without it shrinking or losing its shape] with a steam iron."

With fewer than 200 wet-cleaning establishments across the United States, Milch admits, it is difficult for consumers to find one in their neighborhood. The Environmental Protection Agency has compiled a list of wet cleaners across the country. But customers should not automatically assume that cleaning establishments that bill themselves as "green," "eco-friendly," or "environmentally certified" in the yellow pages use a water-based system. These monikers have been adopted by many dry cleaners who use the new machinery that cuts down on fumes but still uses perchloroethylene.

Simon's consulting company, Business Habits, Inc., of Delray Beach, Florida, advises clients on how to set up professional wet-cleaning operations, either through the use of the new technologies or an updated version of his family's several-stage hand-care process. Scott Seidell, of Ridgefield, Connecticut, is one of Simon's clients. Seidell, an international bond trader, set up his own wet-cleaning shop, The Cleaner Image, in 1995 because he wanted to own a small business and do something for the environment.

Soon after Seidell opened up his shop, two dry-cleaning associations sent him letters in which they threatened to sue him for false advertising, because of his brochure on "environmentally safe cleaning of fine clothing and fabrics." Also, because he did not have a regular laundry machine at his shop, he had planned to contract out cleaning of shirts. But all local shirt-cleaning services, which also rely on dry-cleaning customers, soon refused to take his business. The boycott stopped only after Seidell's lawyer wrote a letter to the dry-cleaning associations, threatening them with antitrust action for illegal restrictions on trade.

"I'm a confident person, and my outlook on business is very global," Seidell says. "I see an opportunity in expanding this, and reaching out to other markets." By not using perchloroethylene, he says, he has reduced insurance costs, workmen's compensation, and special taxes for disposing of hazardous wastes—and also eliminated the need to buy perc itself, at $9 a gallon. He also disputes the notion that wet cleaning is more labor-intensive. Seidell says that as of early 1996, after six months of operation, he had not turned away a single garment because it could not be wet cleaned. "Admittedly, we're going to find something we're not going to want to touch," he said. "I'd almost guarantee you it'll be the same type of thing that they [dry cleaners] are not going to want to touch."

In 1992, the Environmental Protection Agency set out to compare the relative cost and quality of wet and dry cleaning. Its study concluded that Simon's "multi-process wet-cleaning" was, for most dry cleaners, both technically feasible and economically competitive.[8]

Nonetheless, the dry-cleaning and solvent industries generally insist that wet cleaning is not a viable alternative to perc. Mary Scalca, a lobbyist for the International Fabricare Institute, compares it to the powdery aerosol spray that flopped as an instant shampoo in the 1970s. Marshall Miller, a partner in the Washington law firm of Baise & Miller, which has some clients in the dry-cleaning industry, likes to tell this story about the EPA's test cleaning program: "This may be apocryphal," he said, but former EPA Administrator William Reilly's Giorgio Armani suits "came back looking like Little Lord Fauntleroy's suits." EPA officials who ran the testing program say that the anecdote is untrue.

Others in the dry-cleaning industry do not disparage wet cleaning so blithely. Steve Risotto of the chemical industry's Center for Emissions Control says that machine wet cleaning "clearly seems to have a role." But he adds: "Certain fabrics commonly worn and cleaned just don't react well in water. With a wool suit, you're going to have a problem. By controlling temperature and agitation, you can control distortion, or you can restore it by stretching it out, but it's not good long-term. That's where we've come to certain practical limits on the process." To satisfy the needs of all their customers, Risotto argues, cleaners need to offer both wet and dry cleaning. However, the EPA's economic analysis shows that owners of small shops might find it prohibitively expensive to operate both systems.

Risotto uses the example of decaffeinated coffee to make the point that the market will dictate alternatives. Until a little more than a decade ago, manufacturers would decaffeinate their coffees with methylene chloride, the most popular paint stripper on the market (and also a product made by the companies that make perc). When questions were raised about methylene chloride's toxicity, the entire coffee industry turned to water processing for decaffeination, and a popular use for a chlorinated solvent quickly disappeared. "We know from experience, frankly, that if dry cleaners found another solvent they could use economically," Risotto says, "they're going to use it."

Manfred Wentz, of R. R. Street & Company, the distributor of perc for Vulcan Materials, has a highly unusual history in the industry—a vantage point from which he already sees the outlines of a future without chemical dry cleaning.

As an academic—he was for years the chairman of the textile science department at the University of Wisconsin (Madison)—Wentz has seen his industry cope with environmental change before. He had been an early advocate of removing phosphates from commercial detergents, one of the federal efforts in the 1970s to improve the quality of the nation's lakes and streams. Although phosphates had unique cleaning abilities, Wentz says, in the end, they simply were unnecessary. But the detergent industry fought the change vigorously until a few highly successful phosphate-free detergents came on the market, and the clear consumer preference for nonpolluting products drove phosphates off the super-market shelves before federal regulators could.

Wentz states firmly that he does not believe there is evidence that perc is carcinogenic to humans. Nevertheless, he feels that the industry must seek out alternatives. He believes that the federal government has not made enough of an investment in developing or promoting alternatives to dry cleaning, and he points to the example of his native Germany, where the government tightly regulates perc but also gives substantial grants to the industry to develop alternatives. The industry's Institute Hohenstein, in Boennigheim, Germany, is now embarking on a multi-million-dollar project, with government support, to explore replacing perc with carbon dioxide for some uses.

"I think what we have to have here is an open mind," Wentz says. "Namely, we have to indeed look behind dry cleaning with perchloroethylene and look conceptually at what the process is all about. It is a medium of transport for soil away from the garments. And any other medium which fulfills this should be looked at seriously, from the point of view of economics, and from the point of view of technology. Does it clean clothes satisfactorily and does it satisfy the needs of the consumers?"

For perc, a chemical long recognized as dangerous, even greater pressures are building against its use than those that drove phosphates off the market. Even in the early 1970s, when Wentz was a young research director at the International Fabricare Institute in Silver Spring, Maryland, the organization had learned of two deaths in the United States of workers who did not know enough to avoid intense perc vapors. Wentz himself had been overcome by perc, becoming "drunk" and dizzy on its fumes. He began a program of testing dry-cleaning establishments in and around Washington for perc vapors, with an eye to teaching them better maintenance and ventilation practices.

Today, cleaning machinery is far more sophisticated, and dry cleaners use far less perc. But the health and environmental news on perc has gotten much worse; there is evidence of carcinogenicity, reproductive problems, neurological damage. And because federal law now has strict requirements that polluters pay to clean up the toxic wastes of the past, even though they did not know they were polluting at the time they dumped or spilled the chemicals, all dry cleaners face the threat of paying for costly groundwater contamination problems near their shops, just like those faced by the International Fabricare Institute itself.

For so many miracle products of the past that have been revealed as health and environmental hazards in the past 30 years—asbestos, DDT, PCBs, chlorofluorocarbons, and others—it was only the specter of such mounting costs that forced industry to give up its relentless defense and remove its dangerous products from the marketplace.

"In the long run," Wentz says, "the liabilities are getting so high for the solvent that if there is an alternative functioning and available, I guarantee it will take place."

9

Fixing the System

"First, do no harm." This simple idea has lost none of its power since Hippocrates made it the heart of his oath for physicians 2,500 years ago. Today, it is still the guiding principle of public health. Before prescribing a drug or cutting a patient open or ordering a blood transfusion, a doctor must always weigh the potential risks of intervention. If there is reasonable uncertainty about whether doing something is really better than doing nothing, the doctor must do nothing. The first obligation of a physician is to do no harm.

A sort of anti-Hippocratic oath holds sway, however, with toxic chemicals and how government regulates them. Instead of protecting public health, the first obligation of the U.S. regulatory system is to keep chemicals on the market—despite the risks, the costs to individuals and society, and the availability of cheaper and safer alternatives.

Bluntly stated, the system has failed to adequately protect Americans from the hazards of toxic chemicals. Manufacturers can introduce synthetic chemicals into the marketplace without any obligation to employ a precautionary approach that, in the face of uncertainty, puts the interests of public health first. Synthetic chemicals routinely are sold before their effects on people and the environment are well understood. When information surfaces that a chemical is dangerous—a series of adverse reactions, say, or a new study—the burden is on victims, the government, and the public to prove, using information generated and controlled by the manufacturers, that it should not be sold. And when there is uncertainty about a chemical's risks—and there always is—the manufacturer's interests usually take precedence over the public's.

Most of us trust that if a chemical is allowed to be sold in the United States, it must be safe. After all, isn't that why the Environmental Protection Agency and the Food and Drug Administration exist—to ensure that products are safe before they reach the market?

As the stories of alachlor, atrazine, formaldehyde, and perchloroethylene show, this supposed bulwark of government protection has often been an illusion. The EPA rarely tests chemicals for safety. Because of the way the law is written (with the helping hand of chemical industry lobbyists), the government instead relies on manufacturers to test the safety of their own products.

The scandals of Industrial Bio-Test Laboratories in the 1970s and Craven Laboratories in the 1990s show the most blatant hazards of this fox-guarding-the-henhouse system: fabrication of data, skewed results, sophisticated fakery—all meant to show that chemical products are safer and more effective than they really are.

But the most pervasive—and effective—misuse of science by the industry is far less obvious. It happens when formaldehyde manufacturers coordinate a rat study to "test" the accuracy of a prior study that showed the chemical to be carcinogenic, and this time they fix the dose so low and the term so brief that formaldehyde seems safe. It happens when grants from pesticide manufacturers to cash-starved researchers transform the academic discipline of weed science into herbicide science, where the goal is finding new markets for chemicals instad of making agriculture sustainable in the long term for farmers, consumers, and the environment.

It happens, perhaps most subtly of all, in industry's unrelenting efforts to push academic and government researchers into more and more studies of how chemicals cause cancer on the biomolecular level—studies not meant to help cure cancer, or even prevent it, but to exonerate a few chemicals that already have been shown to be carcinogenic. While millions and millions of dollars are poured into the effort to save the market for those chemicals, most chemicals have never even been tested to see if they pose a risk. And the National Toxicology Program, run by the National Institute of Environmental Health Sciences, tests fewer chemicals for their carcinogenicity each year; it has data on the dangers of only 10 to 20 percent of the 70,000 chemicals on the market.

Where are the government watchdogs? They are twisted up in their own leashes. At the big roundtable meetings that EPA officials organize to decide what to do about chemical risks, agency officials seek the "ideas" of manufacturers instead of taking decisive action. To provide "balance," the agency invites a few environmental activists or academics to speak for the public. Instead of acting as a vigorous public advocate, as it is charged to do by law, the EPA has reduced its role to that of a mere mediator of interest groups.

Worried about formaldehyde fumes from wood products, the EPA allows the National Particleboard Association to run the testing program. In the process, tests are never completed on the greatest concern of all— the level of formaldehyde gas that accumulates in mobile homes.

Anxious about atrazine's pervasiveness in drinking water, the EPA accepts Ciba-Geigy's proposal to classify the weed-killer as a restricted-use pesticide and cuts the maximum rate at which farmers can apply it. But more than 90 percent of atrazine users were already qualified to apply restricted-use pesticides, and most were already using atrazine at rates below the new top dosage. The agency's action merely formalizes the agricultural practices that caused the contamination in the first place.

Concerned about perc fumes from dry-cleaning establishments, the EPA agrees with the chemical industry's suggestion that small shops be encouraged to buy state-of-the-art equipment. Dry cleaners are forced to make investments that wed them to perc even while less dangerous alternatives appear on the horizon.

Meanwhile, across the landscape, victims of this regulatory system gone awry fight their solitary battles. Like the Pinkertons in Missouri, the Grahams in Vermont, or the Harrisons in Texas, they become ill without even knowing that they were at risk. If they discover, by accident or persistence, that a chemical product is to blame, their fight has only just begun.

Many lawyers will not take their cases. Because victims typically do not have the money to take a case to trial, most lawyers will agree to represent someone with a chemical injury only if there is some assurance of victory—and a payday—in the end. But there is never an assurance of victory in a toxic-exposure case, because a victim's symptoms can almost always be blamed on something else. And if a lawyer does take the case, the courts require him or her to dig out evidence of corporate misdoings, from the bowels of an industry that has been paying lawyers for years to make sure that those documents never see the light of day.

Getting into the courthouse is an even bigger hurdle. Ironically, the limp protections of the federal regulatory system make it very difficult for victims to find protection—or even compensation—from the court system. Judges have tended to rule that if there is a regulatory law on the books, the responsibility for protecting consumers lies not with the courts, but with the regulators who are responsible for carrying out the law. Routinely, lawsuits are thrown out as preempted, or trumped, because a federal system of protection already exists.

What is the answer? How can a system that is so skewed be righted?

There is no single, all-encompassing solution that will repair a regulatory system that is flawed in so many ways. But a series of practical reforms could, if employed collectively, make the system much fairer than it is now. Indeed, some of these reforms have already proven their merit in states and local communities that have adopted them.

They all reflect a philosophy of openness and market competition. Fixing the system does not have to mean raising taxes, driving up the cost of living, or depriving Americans of the comforts to which they have grown accustomed—to name three arguments that are always raised by industry when changes are suggested. The solutions have been there all along, in a return to the common sense and simple virtue embodied in the Hippocratic oath.

Truth in Testing

Nothing more flagrantly illustrates the shortcomings of the federal regulatory system than the way in which the EPA handles safety testing. When the EPA considers whether or not to approve or ban a toxic chemical, its scientists and bureaucrats do their work in offices, not laboratories. Instead of conducting safety tests themselves, they almost always merely evaluate tests that have been designed and conducted by the chemical's manufacturer or by testing firms that the manufacturer has hired. It is a system tailor-made for manipulation, or even outright fraud.

The EPA sets detailed test protocols, running hundreds of pages, that chemical manufacturers are supposed to follow, but it has little ability to ensure that its rules are being obeyed. The EPA's laboratory inspection program is puny by any reasonable standard—only about 450 of the 2,000 or so labs that do pesticide testing have ever been checked— and the agency has audited only about 3.5 percent of the hundreds of thousands of studies that have been submitted to it. When the EPA does find a problem, the penalty is likely to be a slap on the wrist. In its entire history, the agency has fined only 10 laboratories, and the average fine— $14,360—hardly seems to constitute a deterrent.

Safety testing is not cheap. A standard cancer study, in which a few hundred mice or rats are fed a chemical and watched for two years, can cost $2 million or more. And the two-year rodent study is just one of dozens of tests that are routinely required by the EPA. In the case of a pesticide, for example, the EPA may require 100 or more tests, depending on the product's planned use. The number of tests can climb even higher if evidence of health hazards emerges. For atrazine, one of the most tested of all pesticides, Ciba-Geigy delivered 92 volumes of test results to the EPA in 1995. Since 1983, the company says it has conducted more than 350 tests, at a total cost of $25 million, on atrazene and simazene, a closely related weed-killer.

With 70,000 chemicals in commerce, and hundreds under active review at any given time, it is not realistic to expect that the EPA can take over the job of testing all of them for safety. Turning everything over to government may not be the best solution anyway; to ensure objectivity, the people who evaluate the results of safety tests should not be the same people who conduct the tests.

But surely the current system, in which chemical manufacturers or the laboratories that they hire conduct tests with virtually no oversight from the government, can be improved. At a minimum, the EPA needs to devote more of its resources to inspecting the nation's testing laboratories and auditing the studies that they conduct. More fundamental change may be in order. One option would be to turn all safety testing over to private laboratories that are selected by the EPA instead of by manufacturers, whose role could be limited to picking up the tab for testing their products.

Stop Silencing the Victims

Originally, secrecy orders in lawsuits involving corporations were meant to protect product formulas and other trade secrets that everyone agrees should remain confidential. Today, such orders are routinely used not to protect companies from their competitors but to protect them from more lawsuits over the same faulty or toxic product. To settle their lawsuits and avoid years in the courtroom, plaintiffs often have little choice but to agree to keep secret the details of their cases. Any internal company documents obtained by their lawyers stay sealed, too, out of the sight of other victims and their lawyers.

Judges already have the power to stop this brand of extortion, according to Arthur Bryant, the executive director of Trial Lawyers for Public Justice. Under the federal rules of civil procedure (versions of which are adopted in every state), judges are not supposed to sign secrecy orders—even if the lawyers for both sides agree to such a deal—unless there is "good cause."

"Far too many judges are not being vigilant," Bryant, whose organization works to unseal records in cases involving risks to public health, says. The Trial Lawyers for Public Justice's Project Access got federal appeals courts to order the unsealing of documents showing that executives of Honda Motor Company executives knew about the dangers of all-terrain vehicles before the company marketed them and that Dow Corning Corporation had suppressed evidence of the dangers of silicone breast implants. In both cases, judges had agreed to secrecy orders at the joint request of lawyers for the defendants and plaintiffs, but Bryant's organization convinced the higher courts to rule that there had been no "good cause" to keep the information secret.[3]

The "good cause" standard, however, is itself in danger. Due in part to the efforts of Ciba-Geigy, Dow, Hoechst Celanese, Monsanto, Vulcan Materials, and about a hundred other corporations that belong to a group called the Product Liability Advisory Council, a committee of federal judges has been weighing the idea of throwing out the seldom-invoked language.[4] The manufacturers are urging the courts to get rid of the "good cause" standard as part of an effort to "streamline" the judicial process.[5] In a February 29, 1996, statement to the U.S. Judicial Conference's Committee on Rules of Practice and Procedure, made up of federal judges, the council argued that the change merely puts on paper what already is happening in the courts, permitting "protective" orders at the outset of a suit as long as both parties agree. "Busy federal judges simply do not have the time—or the inclination—to painstakingly oversee the production of documents," it said. The council argued that secrecy orders could still be challenged and that the parties who want them would have to show "good cause" to keep them in place.[6]

However, the proposed new rules require that before a judge lifts or changes a secrecy order, he or she must consider how disruptive such a move would be for any parties that had been relying on the gag order. "It will be virtually impossible for the public to gain access to materials even in the absence of a legitimate reason for secrecy," the Alliance for Justice, a coalition of civil rights and consumer groups, says in its comments opposing the proposed change in rules.[7]

In 1996, the U.S. Judicial Conference began a long-term study on this and other proposals to limit the documents that corporations must disclose when they are sued. After its Committee on Rules of Practice and Procedure decides whether to change the rules that govern secrecy orders, Congress will have six months to vote on whether to go along with its recommendation or reject it and endorse once again the seldom-honored principle that secrecy orders should be the exception, not the rule.

"The entire reason we have courts in this country is to make sure that disputes are resolved in accordance with public notions of justice," Bryant says. "That's why we're paying for courts. [Secrecy orders] prevent the public from finding out the outcome of the cases and whether the result is fair or isn't."

This is true even at a time when Congress is vigorously debating what is happening in the nation's courts—whether there are too many lawsuits, too many big verdicts, too many lawyers. "There are all of these assertions about what is happening in the courts," Bryant says, "and the truth is that no one has any idea what's happening in the courts." No one knows how many lawsuits are settled in secret, but it is thought to be on the order of 90 percent.

"Secrecy undermines the very notion of democratic control of content of the law," Bryant says. "It's impossible for the people to evaluate whether the law, as it is currently being applied, is producing justice."

Don't Let a Label Be a Shield

The federal government's regulatory bureaucracy may be the best friend the chemical industry ever had. Indeed, as scientific advances bring new health risks to light, government agencies have become a crucial protector of manufacturers by propping up dangerous chemicals that otherwise would have been withdrawn under a barrage of victims' lawsuits.

Here is why: Laws such as the Federal Insecticide, Fungicide and Rodenticide Act, the most important pesticide statute, are fundamentally about labeling. After the EPA reviews a pesticide, it approves a label that specifies maximum and recommended dosages and acceptable uses, as well as safety precautions. But a label is only as good as the information on which it is based, and almost all of that information is generated by industry. The safety precautions on the label are typically based on the risks a healthy adult male would face, lending a false sense of security to anyone else who might use the product. Moreover, many years can pass before the agency gets around to updating a label, even if new information about chemical risks has surfaced in the meantime.

When brandished by industry, an EPA label can be a powerful weapon. State and local governments are forbidden under FIFRA from requiring special warnings for pesticides that have an EPA label, and the courts are tightly circumscribed. In recent years, judges have repeatedly ruled that once the EPA has approved a label for a toxic chemical, manufacturers generally cannot be held liable for damage that the chemical might someday cause.

Label-as-shield laws are a tribute to the success of the tobacco industry, which gained important protection against lawsuits as a side benefit of the Surgeon General's warning for cigarettes in the 1960s. Since then, label-as-shield laws have popped up in the oddest places, as when the Department of Housing and Urban Development, through its concern over fumes in mobile homes, protected the formaldehyde industry from lawsuits. Under recent court rulings, federal labeling requirements preempt some of the most important claims of plaintiffs in product-liability lawsuits—including failure to warn, negligence, design defect, and breach of warranty.

Stop Subsidizing Manufacturers

Academic freedom means that, within the bounds of generally accepted ethical principles, no one should tell scientists what they can and cannot research. In practice, however, industry's big money tells scientists what to do every day. Far too many of the nation's scientific minds and resources are devoted to justifying the use of toxic chemicals instead of protecting the public welfare. Consequently, "weed science" researchers, to cite just one example, concentrate on developing new chemical products—and justifying existing products—instead of seeking out nonchemical alternatives.

Short of a revolution at the nation's major research universities, the chemical industry's dominance of the research on its products is unlikely to change. But there is no reason that taxpayer money should be used to promote the already well-heeled chemical industry, which is exactly what happens, for example, when the USDA's Cooperative Extension Service makes pesticides the linchpin of its farmer education programs.

Let the Sun Shine In

Information is power. No one understands that truism better than chemical manufacturers. Their awesome ability to control the direction of scientific research, and the flow of information to the public, has been crucial to the industry's success. Any serious attempt to restore balance to the regulatory system has to involve breaking down the industry's information monopoly.

By focusing on restoring citizen access to information, a few lawmakers have started to make small but real progress in protecting the public without sacrificing industry's right to operate in the free market-place. One of the best-known examples is the 1987 Emergency Preparedness and Community Right-to-Know Act, which was passed in the wake of Union Carbide Corporation's deadly gas disaster in Bhopal, India, and another chemical spill in Institute, West Virginia. It requires every manufacturing facility in the United States to make public an "inventory" of the toxic chemicals it releases from its smokestacks and drainpipes each year.[8]

It is not a perfect system. Some dangerous chemicals are not considered "toxic" under the 1987 law, and the chemical industry was battling the Clinton administration in federal court in 1996 to list additional toxic products. Still, by accessing an EPA computer database known as the Toxics Release Inventory, anyone in the nation can tap into information about chemicals flowing into the air, water, and land in any community.

The same principle of public disclosure could be extended to other areas of chemical regulation. Pesticide manufacturers, for example, are not obligated to disclose the so-called inert ingredients in their products. Those ingredients—which often account for more than 95 percent of a product—are classified as inert only because they do not kill the particular weeds or pests that the pesticide was designed to attack. That certainly does not mean they are safe if consumed by people, whether they are in tainted drinking waters or in the minute residues on the food they eat. Indeed, the EPA has said that many common inert ingredients are highly toxic and suspected carcinogens, including tetrachloride, chloroform, and trichloroethylene. But there is no way to know what is really in a rose spray, a weed-killer, or any other pesticide product unless federal law is changed to compel the disclosure of inert ingredients.

Stronger lobbying disclosure rules could also have profound impact on the regulatory process by exposing the industry's influence-buying efforts to greater public scrutiny. The lobbying-disclosure law that President Clinton signed in 1996 puts tighter controls on gifts and junkets but is likely, many insiders say, to have the effect of shifting industry's surreptitious efforts to other areas, such as phone banking and letter writing by "grass-roots groups" that are actually industry fronts. Richard Gross, a former lawyer for the Consumer Product Safety Commission who battled the formaldehyde manufacturers, has come to believe that the system is distorted by the pressure that Capitol Hill lawmakers bring to bear on the EPA and other regulatory agencies for the interests that help bankroll their campaigns. As long as regulatory decisions are influenced so strongly by campaign money, he says, protecting the public health will stay in the back seat.

Consider the Alternatives

Risk assessment is the chemical industry's favorite catchphrase. In essence, risk assessment is a four-step process in which scientists identify a chemical's hazards, determine the dosages that cause those hazards, assess the degree to which people are exposed to the chemical, and then determine the numeric risk to the population—say, one extra case of cancer per million people. But there is a crucial missing element to risk assessment as it is typically practiced in the United States: the search for alternatives. Risk assessments simply do not consider the fact that even a relatively remote risk should not be tolerated if it can be reasonably avoided.

Similarly, the cost-benefit analyses that the EPA must perform before taking action against a pesticide usually do not take into account nonchemical alternatives. When the agency has to make a decision about whether alachlor should be banned, for example, it considers the effectiveness and price of competing weed-killers but not of nonchemical alternatives that could provide farmers with similar yields at lower costs without adverse health and environmental consequences. Instead, the agency simply comes up with numeric estimates of the chemical's health risks and of the economic impact of taking it off the market.

The numbers themselves are often misleading, because they incorporate what is often little more than guesswork about the number of people exposed to a chemical and the financial impact of its removal. "This whole risk-assessment process is not science, it's voodoo stuff," says James Huff, a senior researcher at the National Institute of Environmental Health Sciences. "You have to include economics, and you take this experimental work that is crude at best, and you come up with a number to 10 decimal places. Just look at the standard-setting for various countries regarding dioxin. Europe and Canada and the United States are miles apart as to what we think is the safe level for dioxin, and we're all supposed to be using risk assessment."

Instead of relying only on a data-driven technique that promises objectivity but fails to deliver, the EPA and other regulatory agencies could examine nonchemical alternatives in addition to guesstimating a chemical's risks and benefits. "The EPA does not see its mission as moving farmers off the pesticide treadmill—it sees its mission as registering pesticides," says Jay Feldman, the national coordinator of the National Coalition Against the Misuse of Pesticides. "The mission should be to register pesticides and reduce inputs while meeting the goals of the pesticide manager, whether it's a farmer or a janitor in a school trying to protect children."

Nicholas Ashford, a public health specialist at the Massachusetts Institute of Technology, says that citizens should demand that the scientists they pay with their tax dollars devote their energies to finding ways to reduce the nation's dependence on synthetic chemicals. "As an intellectual matter, it's interesting to muse about white mice and white rats and why they don't get cancer under certain conditions and if they're bred a certain way," he says. "But it's a distortion of the utilization of public resources to be indulging in intellectual curiosity. Why spend money on this kind of science at all? Why not spend it on technology? If something looks suspicious, why try to save the market for that chemical? Why not find alternatives?"

Get Off the Science-Go-Round

By its very nature, scientific inquiry is something that never ends. Just about any experimental study written up in a reputable scientific journal concludes with a recommendation for further research. Reading that atrazine causes breast tumors in laboratory rats, a toxicologist can quickly rattle off dozens of ways to follow up that study—all of them scientifically legitimate and interesting. The experiment could be repeated for verification, or the doses of the chemical could be altered in an infinite number of ways. A different strain of rat, or a different animal entirely, could be tested. The study's length could be extended or shortened. The chemical could be analyzed in a test tube to see if it damages DNA or stimulates the growth of hormone-sensitive cells. The possibilities are endless, and they all hold out the promise of telling scientists something new about atrazine.

In toxicology, the temptation to perform study after study is particularly strong because the most important question—what are the health risks to humans?—is one that simply cannot be answered directly, because experimentation on humans is unthinkable.

The question is, what happens in the meantime? The endless pursuit of scientific knowledge can be dangerous in a regulatory system in which toxic chemicals are deemed innocent until proven guilty. As scientists work to refine their knowledge of atrazine's health effects, for example, Ciba-Geigy sells tens of millions of pounds of the product every year. The EPA spends a decade arguing with the wood-products industry about how to study formaldehyde fumes in homes—how many houses? how big? where? what temperature?—while Americans continue to buy and remodel homes and be sickened by them, with no clue that their government already knows, but does not inform them, of the risks. Researchers at the National Institutes of Health scrutinize the fine points of how animal tests relate to humans on chemicals such as dioxin (a two-year, multimillion-dollar effort), while they know all along that the dangers are there; it's a matter of degree.

Manufacturers have learned that by focusing on scientific ambiguities, and pointing out that there are so many questions still unanswered, they can all but paralyze federal regulators, who always seem to be worried that they do not know enough. "These are complicated decisions, and we have the scantest information to go by," Reto Engler, a longtime EPA pesticide regulator who is now a consultant, says. "There's such a dearth of information. You're just looking at the tip of the iceberg and you have to try to figure out what the rest of the iceberg looks like."

The industry frequently argues that animal studies are not an accurate indicator of the effects in humans, or that they do not adequately explain the biological mechanism by which a chemical exposure caused a particular health problem. (When it suits their purposes, however, as when they cannot show precisely how a pesticide kills its target insect, manufacturers sometimes claim that evidence of the mode of action is not needed.) James Huff, a toxicologist with the National Institute of Environmental Health Sciences, has heard both arguments again and again. "Everybody wants to discover a mechanism of how something causes cancer," he says. "Well, the truth be known, we haven't discovered one yet. We know a lot about how chemicals cause cancer, but we don't have the complete mechanism for anything. Chemical carcinogenesis was discovered in 1914, and we still don't know exactly how it happens.... The frustrating part is when you have to keep fighting the value of [animal] experimental data. Those who have a vested interest say that until you work on humans, the data are irrelevant. That is a continuing frustration, when in fact much of science and medical practice is based on experimental data."

Engler agrees: "Until we have a better mousetrap—no pun intended—these [animal studies] are still valid studies."

Lawmakers and regulators need to feel comfortable applying their own judgment and acting to protect public health even in the face of scientific uncertainty. In most cases, government officials already have the legal authority to take action, yet almost never use it—in large part because industry can always find questions that need further research. Lawsuits, the remedy of last resort, are poor substitutes for prevention-oriented regulation. "The best way to deal with these kind of chemicals is not through torts but through good science which shows there are risks and then translates that into public action without requiring people to die or be maimed first," says Dr. J. Routt Reigart, a professor of pediatrics at the Medical University of South Carolina and the former chief of the committee on environmental hazards of the American Academy of Pediatrics. Adds Huff: "Someone said that prevention delayed is prevention denied. I believe that."

Arm Consumers With Facts

In a free society, is there any way for the government to advance the precautionary principle on behalf of its citizen-consumers? The chemical industry's message is no; the free market should determine what products will be produced, sold, and bought. The big problem, however, is that consumers are denied the kind of information that gives them the power to speak, through their pocketbooks, in the free marketplace.

The Toxics Release Inventory provides some information about the kinds of chemicals that factories and other big polluters are putting in the air, water, and soil. But what about the chemicals being released in bedrooms, kitchens, and nurseries? There has been no Bhopal disaster of the home to spur lawmakers, only the quiet suffering of uncounted families who, in many cases, do not even know what is happening to them.

California, in a 1986 law that was approved two-to-one directly by the state's voters, appears to have hit on a novel idea for regulating toxic chemicals. In half the time that it has taken the EPA to set standards for fewer than a dozen hazardous air pollutants, California has identified 500 chemicals as toxic and has adopted clear, numerical limits for 282 chemicals for which there is evidence of cancer or birth-defect risks.[9]

Moreover, the state has succeeded without provoking a single lawsuit from industry or environmental organizations. Industry originally decried Proposition 65, the Safe Drinking Water and Toxic Enforcement Act, known commonly as the Green Labeling Law, as unrealistic and bound to produce chaos. Now, Republican governor Pete Wilson's administration has described the amount and quality of the scientific risk assessment as "100 years of progress" compared to the federal government's.

The secret of the law's success is that, first, it does away with the years of scientific debate and sets unambiguous limits for toxic chemicals by relying on data collected with the use of taxpayer money by the National Toxicology Program (in the Carcinogenic Potency Database at Lawrence Berkeley Laboratories.) No endless studies of how a chemical causes cancer are considered. The fact that there is a risk is all that is needed for action.

Second, and more important, the California law puts information in the hands of consumers. To protect the public, the law relies on the very marketplace incentives that the industry lauds. The law does not ban any chemicals or limit their use in any way. It simply says that if a product contains more than the state-specified limit of any chemical on the list, the manufacturer must clearly display that information on the label.

For toxic chemicals for which no limit has been set, all products containing even minute amounts of the chemical must be labeled. It is a dramatic shift of the burden of proof to manufacturers; as soon as there is evidence of risk, a chemical is guilty until proven innocent. As a result, manufacturers clamor to have standards set for their chemicals. And they have an incentive, for the first time, to remove toxic chemicals from products.

Clearly, there are trade-offs in the law that benefit industry. The California law allows consumer products to contain low levels of the 282 chemicals for which standards have been set. And the law's emphasis on informing consumers instead of banning products clearly throws the ball into the uncertain court of the marketplace.

But consider the results. In the past 10 years, largely because of the California law, perchloroethylene has been removed from a host of household products for cleaning spots from carpets, upholstery, and clothing—the best known of which is K2R, the spot remover manufactured by Dow, the nation's leading perc manufacturer. Gillette took its Liquid Paper correction fluid off the market, then brought it back as "New Improved" Liquid Paper, without trichloroethylene, a solvent that is a close cousin of perc. Manufacturers of fine china—among them Lenox, Tiffany, and Wedgewood—have eliminated lead from their products rather than affix a yellow label that would alert consumers in California to a health hazard that had previously been detected in about half of the china on the market. Fourteen manufacturers of ground well-water pumps all agreed in 1994 to remove leaded brass from their products, which could release 100 to 1,500 parts per billion lead, when the maximum amount is one-fourth of a part per billion. They knew how to do this, but never bothered until Prop 65 became law. And because California represents at least 15 percent of the U.S. marketplace, manufacturers have reformulated these products as they are sold all around the country—not just in the Golden State.

"All conventional [toxics] laws were built to reward delay and penalize progress," says David Roe of the Environmental Defense Fund, the author of California's Proposition 65. "It's worth it to hire large teams of gentlemen in white lab coats to argue out every esoteric twist in the risk assessment debate: Are rats the same as mice? If they fall over to the left, is that the same as falling over to the right? Are Swedish rats the same as Norwegian rats? The conventional system invites endless debate over tiny points because endless debate stalls anything. You can look at this as not bad faith on the part of business, but a rational response to the incentives that Congress built into the conventional laws.

"What California did was a conscious experiment. What if we changed those incentives, and rewarded progress and penalized delay? It's very simple—protection of the public will go into effect whether the homework is finished or not. You can't stall the law by stalling the homework. Indeed, the law is more stringent if the homework isn't done than if the homework is finished. The burden of proof on the scientific details has been shifted to the businesses using the chemicals."

There are weaknesses and omissions in the law. Some of the most dangerous products on the market are not covered by Proposition 65 because the stringent language of FIFRA preempts state regulation. Similarly, HUD's weak labeling regulation for mobile homes supersedes California's stringent labeling standards.

Gilbert Omenn, the dean of the University of Washington's School of Public Health and chairman of the President's Commission on Risk Assessment, supports a national truth-in-labeling law, as does as the Risk Dialogue Group, an industry-government coalition chaired by Lee Thomas, a former EPA administrator (in the Reagan administration) who is now Georgia-Pacific's senior vice president for paper.

The chemical industry, and its lobbyists in the battlefield, do not like the idea. Sara Schotland, an attorney for formaldehyde manufacturers, believes that the chemical limits—the thresholds that trigger a labeling requirement—are too low. Many others argue that there is no benefit to consumers in being bombarded with cancer labels.

"The concept of providing information and sharing information is a very valid concept," says Jerry Dunn, vice president for environmental, health, and safety affairs for Hoechst Celanese. "The problem with labels is that so many are required that when an individual walks up to a drum of formaldehyde, there may be 12 to 15 labels, and people won't have the patience to sit down and read them all. They throw up their hands in frustration."

Whether it is the California approach or another, observers such as Mark Greenwood, the former director of the EPA's Office of Pollution Prevention and Toxics who now is an industry lawyer, believe that it is inevitable that regulatory programs of the future will be designed first and foremost to put information in the hands of consumers. One reason, he points out, is that it is cheaper than traditional regulation. The EPA's huge Clean Air Act and Clean Water Act programs touch only a few hundred manufacturers; the Toxics Release Inventory affects 23,000 industry facilities at a fraction of the cost. Greenwood also notes that there are time-tested models for such information-based regulation; the Securities and Exchange Commission has relied on the principle of public disclosure for 50 years to protect U.S. investors.

Such fundamental change to protect consumers from health threats in their homes and offices would, of course, require dynamic leadership and true democratic pressure. This is especially true in light of industry's history of twisting labeling laws to its advantage, carving out exceptions for favored products, and blunting the edge of environmental regulation. But until the system is changed, the burden of the precautionary principle will remain on consumers, who must somehow make sense of a confusing thicket of misleading information and slick chemical industry advertising in order to protect themselves and their families.

10

The Cost of
Toxic Deception

In spring 1996, Clifford T. "Kip" Howlett worked to enlist a new battalion of foot soldiers in the chemical industry war. This particular battle, however, would not be over formaldehyde, the product he defended in the 1980s with a platoon of mobile-home dealers, scientists, and PR experts.

Howlett had retired as Georgia-Pacific's vice president of government affairs three years earlier, and he owned a considerable block of stock in the company. Now, he was a vice president of the Chlorine Chemistry Council in Alexandria, Virginia, a new outfit that chemical manufacturers had assembled to protect yet another product.

Although chlorine is a cheap, widely used method of fighting bacteria in water systems, its byproducts, a 1992 study estimated, may be to blame for about 10,000 bladder and rectal cancer cases in the United States each year. Also, there are safer alternatives, including chlorine dioxide, ozone, and ultraviolet light. Weighing the known dangers of bacteria against the possible dangers of chlorine, the EPA proposed that households have no more than 4 milligrams per liter of chlorine in their tap water, an easy goal to reach because it is at least double the current average.

The chlorine industry opposed the limit, even though water disinfection accounts for only 1.5 percent of world use of its product. So just as he had encouraged the small businesses that buy formaldehyde products to march with manufacturers against government regulation of that chemical, Howlett sought important and sympathetic chlorine customers to join the fray on behalf of chlorine.

No one was quite sure how, but the Chlorine Chemistry Council obtained the 4,000-name mailing list of the public drinking water companies that are members of the American Water Works Association.[1] The firms, owned by cities, towns, and some private companies, received letters asking them to lobby Congress to curb the EPA's efforts.

At the same time, groups that help people with AIDS received similar letters from the Chlorine Chemistry Council. This time the council argued that the EPA's regulation of chlorine would hurt people with weakened immune systems, who can become fatally ill from bacteria such as cryptosporidium in tap water. The letter did not mention that mere chlorination does not kill cryptosporidium (although treatment with nonchemicals, such as heat or ultraviolet light, does).

In an April 1996 interview, Howlett was unapologetic: "We are proud of the fact that chlorine has saved more lives than any other chemical in the history of mankind," he told a reporter for the Associated Press. As for the effort to enlist AIDS victims and public water systems on behalf of chlorine: "They've got an interest in making sure this rule comes out right."

As it turned out, neither water companies nor the AIDS advocacy groups agreed to take up arms for the chlorine industry, and Congress's revisions in the Save Drinking Water Act, passed in July 1996, did not contain what the Chlorine Chemistry Council wanted. But the skirmish established a framework for future debate on chlorine: policymakers will remember that they should weigh the chemical's risks against its life-saving benefits. And as Howlett undoubtedly knew from his years of work on formaldehyde, chemical wars are won with long-term strategy.

The stories of alachlor, atrazine, formaldehyde, and perchloro-ethylene are not strange deviations from the chemical industry's custom-ary way of doing business in Washington. Slanted science, political pressure, and PR juggernauts are used again and again to keep dangerous but profitable products on the market and to crowd out promising alternatives. These four chemicals have left a richer history and clearer paper trail in their wake than most. But if alachlor, atrazine, formaldehyde, and perc were banned tomorrow, it would not begin to address the real problem: the industrywide pattern of abuse we call toxic deception.

Howlett is not the only player in the stories of these four chemicals who has shown up, playing a similar role, on behalf of other chemical industry products. William Gaffey, one of the Monsanto scientists who led the joint-industry National Cancer Institute study in 1986 that raised doubts about whether formaldehyde caused lung cancer in humans, also was responsible for the 1980 study of workers that seemed to show that Agent Orange, the Vietnam-era herbicide, could not be blamed for cancer.

For years, Monsanto had used Gaffey's work to defend itself from lawsuits by veterans and chemical industry workers who felt that they had been harmed by dioxin. It was not until 1990 that an EPA official discovered (through documents that had surfaced in lawsuits) that Gaffey and his colleagues had classified four workers as "unexposed" to dioxin when the same workers had been classified as "exposed" in a previous study. Analyses of the corrected data showed a statistically significant increase in lung cancer among dioxin-exposed workers.[2] Still, the EPA has been reluctant to regulate dioxin as a carcinogen. Dioxin continues to be a major pollutant and byproduct of production in the wood and paper industry.

In Florida, in June 1996, a Dade County jury, after hearing another tale of toxic deception, awarded nearly $4 million to a West Palm Beach couple whose son has been born without eyes. The boy's mother, Donna Castillo, a schoolteacher, had been drenched by Benlate DF, a fungicide produced by DuPont, when she was walking past a tomato field when six weeks pregnant.[3] Jurors learned of Benlate tests conducted in the 1970s by a DuPont scientist, Robert Staples, that found rats born without eyes after exposure to the chemical. Those tests prompted the EPA to recommend in 1977 that the chemical be marketed with a birth-defect warning for pregnant women. But DuPont successfully fought the battle to keep the warning label off. In memorandums produced at the Dade County trial, DuPont executives ridiculed Staples as the "Lone Ranger." DuPont repeated the tests with a new methodology and found no birth defects, and the EPA dropped its birth-defect warning effort. At the trial, Staples swore under oath that the methodology of his original tests had been sound.[4]

Lawyers for the Castillos say other families with Benlate-caused birth defects were waiting in the wings to file suit against DuPont, and 40 cases are pending in other countries. But indications are that DuPont, which has been fined more than $100 million for withholding evidence in previous Benlate cases (over damage the product did to crops, not the birth defects), does not intend to give up easily.

Lorraine and Kevin Burke of St. James, New York, learned of toxic deception when they had two children within 15 months who had the same rare birth defect: congenital cataracts. Physicians at the Neurological Institute at Columbia-Presbyterian Medical Center in New York City believe that the children were harmed inside the womb by the pesticide the Burkes used, Rid-A-Bug, which contains both the solvent xylene and Dow Chemical Company's insecticide, Dursban. The family had used pesticides because they feared their dog would carry ticks—and Lyme disease—into their home. And they had been so concerned about the health effects of pesticides that they had kept their older daughter out of the home when it was sprayed, as the pesticide label recommended. But the label did not warn that pregnant women should beware of the lingering effects of the substance. The Burkes have a $125 million suit against Dow, which the company has tried to have dismissed on the grounds that the EPA does not require pesticides to carry warnings for pregnant women on their labels.[5]

In California, parents can blame toxic deception for the yearly application of a deadly pesticide on strawberry fields close to thousands of Orange County children who attend nearly 300 schools and licensed day-care centers. Methyl bromide, produced by Great Lakes Chemical Corporation and other companies, has killed 18 people since 1982 and poisoned 454 others, according to the Washington-based Environmental Working Group.[6] In laboratory tests, methyl bromide has caused rabbits to be born without gall bladders and dogs to slam their heads against the sides of their cages. Methyl bromide also is known to deplete the earth's ozone layer.

Safer methods of strawberry fumigation are available. But the industry has successfully fought efforts at the state and federal levels to regulate its use. In a January 1995 internal industry memo obtained by Ozone Action, a Washington-based environmental group, a coordinating body named the Methyl Bromide Working Group, warns chemical manufacturers to "continue to work together" to be able to use methyl bromide "well beyond the year 2001," when use of the chemical is supposed to be phased out under an international treaty to protect the ozone layer.

"Be very careful what you say publicly about methyl bromide alternatives," the memo says. It goes on to say that if word about the emerging alternatives leaks out, environmentalists will "try to prevent you from using methyl bromide ever again."[7]

If you don't live near the orchards of Florida or California, is there any reason to worry? The main ingredient in the most popular insect repellents on the market is the pesticide N,N-diethyl-m-toluamide, or Deet. It can cause headache, restlessness, irritability, crying spells, and behavioral changes in children in mild doses. Severe poisoning by the product—which is increasingly sold in 100 percent concentrations—can cause slurred speech, tremors, convulsions, and coma.

The Journal of the American Mosquito Control Association has published a litany of the admittedly rare but devastating ills that Deet has caused. A 17-month-old girl suffered brain irritation and died after frequent use of 20 percent Deet. A three-year-old girl who had used 15 percent Deet heavily had convulsions and brain irritation. An eight-year-old boy had convulsions after Deet was sprayed on a carpet. A five-year-old girl had convulsions and died after heavy use of 10 percent Deet.[8]

Still, the Chemical Specialties Manufacturers Association quashed an effort by New York state to regulate the product and possibly ban 100 percent Deet solution. In 1990, the EPA launched a reevaluation of the chemical, but in 1996 the agency still was sorting through more than 50 new and old scientific studies, all provided by the manufacturers, to decide what to do.

Most people want to decide for themselves what risks are worth taking. They would like every bit of information possible to protect their children and themselves from harm. They would like the taxes they pay for an Environmental Protection Agency to be used to defend health and safety, not to moderate disputes between environmental organizations and industry.

Instead, citizens are shielded from information more effectively than they are shielded from environmental harms. The government works with industry to prevent confusion and alarm. And the script is the same, whether written by the Formaldehyde Institute, the American Crop

Protection Association, the Center for Emissions Control, the Chlorine Chemistry Council, or the Methyl Bromide Working Group: There is nothing to worry about, they say.

And mothers continue to die young, farm families are ravaged by illness, children are born without eyes, youngsters lapse into seizures on a carpet, and no one really knows why cancer and infertility rates are rising. Perhaps the worst consequence of the chemical industry's untruths is not any one personal tragedy but the general loss of society's will and drive to find better ways to do the things we have relied on chemicals to do for just these 50 years. Professional clothes-cleaning with water, careful tilling techniques that don't erode the soil, formaldehyde-free home products, the use of ultraviolet light to disinfect water—these bright seeds of a future without harmful chemicals may never receive the nurturing they need to burst forth and change the way we live. If they do not, then no single chemical will ever prove as toxic as deception itself.

Sources

Interviews

The authors interviewed more than 200 people for this book over a two-year period. A few agreed to be interviewed only on the condition that they not be identified or quoted, but the vast majority spoke without any restrictions. Not everyone who was interviewed is identified in the text or footnotes, but all of them helped to shape the book, and they all have the authors' gratitude.

Here is a list of people with whom the authors conducted on-the-record interviews:

Karem Ahmed, former staff scientist, Natural Resources Defense Council

Nicholas Ashford, professor of technology and policy, Center for Technology Policy and Industrial Development, Massachusetts Institute of Technology

Byrd Baggett, former mobile-home owner and plaintiff in a formaldehyde-related injury case, Houston

Joe Bailey, special review manager, Office of Pesticide Programs, Environmental Protection Agency

David Baker, professor of biology and director, Water Quality Laboratory, Heidelberg College

Nicole Ballenger, program officer, National Academy of Sciences Board of Agriculture

Daniel Barolo, director, Office of Pesticide Programs, Environmental Protection Agency

Robin Bellinder, associate professor of horticulture, Cornell University

Charles Benbrook, former director, National Academy of Sciences Board of Agriculture

Robert Bennett, attorney, Bennett, Krenek & Hilder, Houston

Aaron Blair, chief, Occupational Studies Section, National Cancer Institute

Jerry Blondell, epidemiologist, Office of Pesticide Programs, Environmental Protection Agency

Anne Bloom, staff attorney, Trial Lawyers for Public Justice

Kenneth Bogdan, research scientist, New York State Center for Environmental Health

Raymond Booth III, attorney, Jacksonville, Florida

Lynn Marie Bower, codirector, Healthy House Institute, Bloomington, Indiana

H. Leon Bradlow, director, laboratory of biochemical endocrinology, Strang-Cornell Cancer Research Laboratory

Peter Breysse, professor, Department of Environmental Health, University of Washington

Eugene Brooks, attorney, Atlanta

Arthur Bryant, executive director, Trial Lawyers for Public Justice

Susan Burdick, weaver, Yurok Indian Reservation

Orvin Burnside, professor of weed science, University of Minnesota

Frederick Buttel, professor of rural sociology, University of Wisconsin

Denny Caneff, executive director, Wisconsin Rural Development Center

Bill Carpenter, sole practitioner, Portland, Oregon

Keith Cherryholmes, former assistant director, Iowa State Hygienic Laboratory

Karen Clark, attorney-adviser, Office of General Counsel, Environmental Protection Agency

Robert Clark, manager of public relations, Ciba-Geigy Corporation

Harold Coble, professor of weed science, North Carolina State University

Stuart Z. Cohen, former groundwater team leader, Office of Pesticide Programs, Environmental Protection Agency

James Conlon, former deputy director, Office of Pesticide Programs, Environmental Protection Agency

Rory Conolly, manager of formaldehyde program, Chemical Industry Institute of Technology

Bill Craven, attorney, Topeka, Kansas

Kent Crookston, professor of agronomy, University of Minnesota

Kim Cutchins, president, National Peanut Council

Devra Lee Davis, senior fellow, World Resources Institute

Dowell A. Davis, pharmacologist, Food and Drug Administration

James H. Davis, attorney, Los Angeles

David Dull, associate director, Agriculture and Ecosystems Division, Office of Compliance, Environmental Protection Agency

Jerry Dunn, vice president for environmental, health, and safety affairs, Hoechst Celanese Corporation

Venus Eagle, registration manager, Office of Pesticide Programs, Environmental Protection Agency

Bob Ehart, former public affairs officer, Ciba-Geigy Corporation

Reto Engler, former senior science adviser, Office of Pesticide Programs, Environmental Protection Agency

Samuel Epstein, M.D., School of Public Health, University of Illinois Medical Center

Rick Exner, extension specialist, Iowa State University Cooperative Extension

Jay Feldman, executive director, National Coalition Against the Misuse of Pesticides

Penelope Fenner-Crisp, deputy director, Office of Pesticide Programs, Environmental Protection Agency

Richard Feulner, director of regulatory affairs, Ciba-Geigy Corporation

Eric Frumin, health and safety director, UNITE! (formerly the Amalgamated Textile and Garment Workers Union)

George Fuller, director of product registration and regulatory affairs, Monsanto Company

Susan Galliher, program analyst, Budget Division, Environmental Protection Agency

Leonard Gianessi, senior research associate, National Center for Food and Agricultural Policy

Terry Gloriod, president, Missouri River Public Water Supply Association

Thad Godish, professor and director, Indoor Air Research Laboratory, Ball State University

Rebecca Goldburg, senior scientist, Environmental Defense Fund

Patti Goldman, staff attorney, Sierra Club Legal Defense Fund

Don Goolsby, chief, Midcontinent Herbicide Program, U.S. Geological Survey

Howard Goring, supervisor of water and wastewater, City of Higginsville, Missouri

Brigitte Graham, Norton, Vermont

Elizabeth Graham, Norton, Vermont

Sara Greensfelder, executive director, California Indian Basketweaver Association

Mark Greenwood, attorney, Ropes & Gray, Washington, D.C.

Norma Grier, executive director, Northwest Association for Alternatives to Pesticides

Richard Gross, attorney, Rosenman & Colin, Washington, D.C.

Ken Halden, corporate communications director, Georgia-Pacific Corporation

George Hallberg, chief of environmental research, University of Iowa Hygienic Laboratory

Michael Hart, attorney, Columbia, South Carolina
Chuck Hassebrook, program leader, Center for Rural Affairs
William Hausler, Jr., former director, Iowa State Hygienic Laboratory
Bill Hayes, product steward, Dow Chemical Corporation
Rick Hind, legislative director, Toxics Campaign, Greenpeace
Maureen Hinkle, director of agricultural policy, National Audubon Society
Dan Holman, communications manager, Monsanto Company
James Howard, assistant U.S. attorney for Maryland
James Huff, toxicologist, National Institute of Environmental Health Sciences
Nick Iammartino, vice president for public affairs, Borden, Inc.
Diane Ierley, former special review manager, Office of Pesticide Programs,
    Environmental Protection Agency
John Ikerd, associate professor of agricultural economics, University of
    Missouri
Peter Isacson, former professor of epidemiology, University of Iowa
Lyle Jackson, county sanitarian, Fayette County, Iowa
Bruce Jaeger, designated federal official, Science Advisory Panel, Environmen-
    tal Protection Agency
Cyrus Jaffari, president, Caspian, Inc.
Evelyn Janik, Houston, former mobile-home resident and plaintiff in
    formaldehyde-related injury case
Rhonda Janke, associate professor of agronomy, Kansas State University
Jeffrey Johnson, attorney, Dickstein & Shapiro, Washington, D.C.
Dennis Keeney, director, Leopold Center for Sustainable Agriculture, Iowa
    State University
Richard Kelley, program consultant, Iowa State Hygienic Laboratory
Patricia Kenworthy, director of regulatory affairs, Monsanto Company
Ellen Kirrane, professor, Hunter College
Andrew Klein, manager of technology, Monsanto Company
Michael Kramer, assistant professor of pediatric pulmonology and allergy,
    University of Missouri
Philip Landrigan, director of environmental sciences, Mt. Sinai School of
    Medicine
Brock Landry, attorney, Jenner & Block, Washington, D.C.
Sandy Lange, public liaison, National Toxicology Program
Paul Lapsley, director, Regulatory Management Division, Environmental
    Protection Agency
Jack Lauber, New York Department of Health
Evan T. Lawson, attorney, Lawson & Weitzer, Boston
Frances Liem, chief, Laboratory Data Integrity Branch, Office of Pesticide
    Programs, Environmental Protection Agency

Richard Lippes, attorney, Allen, Lippes & Shonn, Buffalo, New York
Adele Logan, vice president, Strat@comm
Randall Lutz, attorney, Smith Somerville & Case, Baltimore
Charles Lynch, medical director, Iowa State Cancer Registry
Vic Madsen, farmer, Audubon, Iowa
Nadav Malin, managing editor, *Environmental Building News*, Brattleboro, Vermont
William Marcus, senior science adviser, Office of Ground Water and Drinking Water, Environmental Protection Agency
Richard Margosian, director, National Particleboard Association
Alex Martin, professor of agronomy, University of Nebraska
John Meyers, former director, Current Research Information System, U.S. Department of Agriculture
Neal Milch, president, Aqua Clean Systems, Inwood, New York
Fred Miller, director, School of Natural Resources, Ohio State University
Marshall Miller, attorney, Washington, D.C.
Roger Mitchell, dean, College of Agriculture, Food, and Natural Resources, University of Missouri
Marion Mlay, former director, Office of Wetlands, Oceans, and Watersheds, Environmental Protection Agency
Alan Moore, technical information specialist, Current Research Information System, U.S. Department of Agriculture
Ron Munger, associate professor of nutrition and food science, Utah State University
Don Nantkes, ethics officer, Office of General Counsel, Environmental Protection Agency
Alex G. Ogg Jr., supervisory plant physiologist, Agricultural Research Service, U.S. Department of Agriculture
Craig Osteen, deputy director, National Agricultural Pesticide Impact Assessment Program, U.S. Department of Agriculture
Kenneth Owen, executive director, Rathbun Regional Water Association
David Ozonoff, chairman, environmental health program, Boston University School of Public Health
Jean Parker, researcher, Carcinogenic Risk Assessment Division, Environmental Protection Agency
Jim Parochetti, chief weed scientist, Cooperative State Research Service, U.S. Department of Agriculture
Arthur Perler, former chief of science and technology, Office of Drinking Water and Ground Water, Environmental Protection Agency
Dave Pimentel, professor of insect ecology and agricultural science, Cornell University

Carl Pray, associate professor of economics, Rutgers University

John Radin, acting national program leader for weed science, Agricultural Research Service, U.S. Department of Agriculture

Dr. David Rall, former director, National Institute of Environmental Health Sciences

Stephen Randtke, professor of civil engineering, University of Kansas

Herb Reed, corporate communications, Hoechst Celanese Corporation

J. Routt Reigart, professor of pediatrics, Medical University of South Carolina

Joseph Reinert, senior policy analyst, Office of Pesticide Programs, Environmental Protection Agency

Katherine Rhyne, attorney, King & Spalding, Washington, D.C.

Bonnie Rice, former campaigner, Greenpeace

Steve Risotto, director, Center for Emissions Control

Patty Ritchie-Blase, environmental specialist, Missouri Department of Natural Resources

J. Alan Roberson, director of regulatory affairs, American Water Works Association

David Roe, senior attorney, Environmental Defense Fund

Avima Ruder, epidemiologist, National Institute of Occupational Safety and Health

Bob Rumpza, managing director, Ceres Communications

Mary Scalca, legislative director, International Fabricare Institute

Sara Schotland, attorney, Cleary, Gottlieb, Steen & Hamilton, Washington, D.C.

Debbie Schprentz, senior research associate, Natural Resources Defense Council

Judith Schreiber, toxicologist, New York Department of Health

Edward Schweizer, plant physiologist, Natural Resources Research Center, U.S. Department of Agriculture

Scott Seidell, owner, The Cleaner Image, Ridgefield, Connecticut

George Semanyiuk, formaldehyde project manager, Chemical Control Division, Office of Pollution Prevention and Toxics, Environmental Protection Agency

Peter Shelley, staff attorney, Conservation Law Foundation

Robert Shields, attorney, Doffermyre, Shields, Canfield, Knowles & Divine, Atlanta

Terry Shistar, adjunct professor of environmental studies, University of Kansas

Richard Simon, president, Business Habits, Inc.

Charles Smith, former director, National Agricultural Pesticide Impact Assessment Program, U.S. Department of Agriculture

Connie Smrecek, former president, Save Us From Formaldehyde Exposure Repercussions
Henk Snow, vice president, Medite Corporation
Carol Sonnenschien, professor of cell biology, Tufts University School of Medicine
Henry Spencer, toxicologist, Office of Pesticide Programs, Environmental Protection Agency
John Stauber, executive director, Center for Media & Democracy; editor, *PR Watch*
Nancy Steorts, consultant, Dallas
Darrell Sumner, manager of health and safety issues, Ciba-Geigy Corporation
James Swenberg, director of biochemical toxicology and pathobiology, Chemical Industry Institute of Toxicology
Kimberly Thompson, fellow, Harvard School of Public Health
Dennis Tierney, manager of environmental fate and effects, Ciba-Giegy Corporation
Thomas Vaughn, epidemiologist, University of Washington
Toby Vigod, former executive director, Canadian Environmental Law Association
R. Frederick Walters, attorney, Walters, Bender & Strohbein, Kansas City
Barbara Warren, director, Consumer Policy Institute, Consumers Union of the United States
Manfred Wentz, R. R. Street & Company
Richard Wiles, vice president for research, Environmental Working Group
Walt Woods, deputy adminstrator for partnerships, U.S. Department of Agriculture
Garth Youngberg, director, Wallace Institute for Alternative Agriculture
Sheila Hoar Zahm, epidemiologist, National Cancer Institute.
Sergio Zori, architect, Sea Cliff, New York

Documents

All internal memorandums of Georgia-Pacific Corporation and all records of meetings of the Formaldehyde Institute were made available to the authors by Robert S. Bennett, a Houston lawyer who obtained the documents in the course of 15 years of litigation on behalf of people exposed to formaldehyde in mobile homes. Bennett made available only those documents that were not subject to secrecy orders.

The authors obtained the memos and correspondence related to the handling of formaldehyde in wood products by the Environmental Protection Agency's Office of Toxic Substances from the EPA's formaldehyde docket and in response to a request for documents under the Freedom of Information Act. All these documents subsequently were made available as Administrative Record 127 in the EPA's Toxic Substances Control Act public docket.

All internal memos and documents of the International Fabricare Institute were obtained from the public record in the case of Westfarm Associates v. International Fabricare Institute, Docket #92-9, U.S. District Court, Baltimore, Maryland.

All records relating to the history of the federal regulations on clothing-care labeling were obtained from the Federal Trade Commission's public record, Docket #215-22, which spans 28 years. The authors obtained most of the memos and correspondence related to the handling of perchloroethylene by the EPA's Office of Toxic Substances in response to requests for documents under the Freedom of Information Act. Several memos requested were withheld or redacted, however, because they contained carcinogenic-risk numbers that the EPA viewed as protected information related to the agency's internal deliberative process. The numbers that the EPA has generated, however, are well known among parties to the regulation, both chemical industry and environmental and labor groups. To provide the public with the complete story of EPA's dealings with the dry-cleaning chemical, the authors obtained these memos through other sources who asked that their identities not be disclosed.

Some memos and correspondence related to regulation of perchloroethylene also were obtained from the EPA's Office of Air and Radiation, National Emissions Standards for Hazardous Air Pollutants (Perchloroethylene) Dockets #88-11 and 93-45.

The authors obtained hundreds of letters, memos, and other internal EPA documents pertaining to alachlor and atrazine from their respective EPA dockets and through the Freedom of Information Act. They are available through the Office of Pesticide Programs public docket in Crystal City, Virginia.

Notes

Abbreviations

CPSC: Consumer Product Safety Commission
EPA: Environmental Protection Agency
FTC: Federal Trade Commission
FDA: Food and Drug Administration
FI: Formaldehyde Institute
GAO: General Accounting Office
IARC: International Agency for Research on Cancer
IFI: International Fabricare Institute
NTP: National Toxicology Program, National Institutes of Health
OMB: Office of Management and Budget

Introduction: The Invisible Threat

1. Pinkerton v. Temple Industries, cv-186-4651-cc, Clay County, Missouri, November 1989.

2. The warning read: "WARNING: IRRITANT. THIS PRODUCT CONTAINS A UREA FORMALDEHYDE RESIN AND MAY RELEASE FORMALDEHYDE VAPORS IN LOW CONCENTRATIONS. FORMALDEHYDE CAN BE IRRITATING TO THE EYES AND UPPER RESPIRATORY SYSTEM, ESPECIALLY IN SUSCEPTIBLE PERSONS SUCH AS THOSE WITH ALLERGIES OR RESPIRATORY AILMENTS. USE WITH ADEQUATE VENTILATION. VENTILATION RATE SHOULD NOT BE LESS THAN ONE (1) AIR CHANGE PER HOUR. IF SYMPTOMS DEVELOP, CONSULT YOUR PHYSICIAN."

3. "IARC & NTP Carcinogen List," compiled by the NTP, http://ntp-server.niehs.nih.gov.

4. Devra Lee Davis, Aaron Blair, and David G. Hoel, "Agricultural Exposures and Cancer Trends in Developed Countries," *Environmental Health Perspectives* 100 (1992): 42–43.

5. Devra Lee Davis, Gregg E. Dinse, and David G. Hoel, "Decreasing

Cardiovascular Disease and Increasing Cancer Among Whites in the United States From 1973 Through 1987," *Journal of the American Medical Association* 271, no. 6 (9 February 1994): 431.

6. Samuel S. Epstein, "Evaluation of the National Cancer Program and Proposed Reforms," *American Journal of Industrial Medicine* 24 (1993): 112.

7. Susan Devesa et al., "Recent Cancer Trends in the United States," *Journal of the National Cancer Institute* 87, no. 3 (1 February 1995): 179–80.

8. Ibid.

9. Howard M. Hayes et al., "Case-Control Study of Canine Malignant Lymphoma: Positive Association With Dog Owner's Use of 2,4-Dichlorophenoxyacetic Acid Herbicides," *Journal of the National Cancer Institute* 83 (4 September 1991): 1226–31.

10. Environmental Working Group, Physicians for Social Responsibility, and the National Campaign for Pesticide Policy Reform, *Pesticide Industry Propaganda: The Real Story* (Washington: Environmental Working Group, 1995), 6.

11. Jack K. Leiss and David A. Savitz, "Home Pesticide Use and Childhood Cancer: A Case-Control Study," *American Journal of Public Health* 85, no. 2 (February 1995): 249–52.

12. Aaron Blair and Shelia Hoar Zahm, "Cancer Among Farmers," *Occupational Medicine: State of the Art Reviews* 6, no. 3 (July–September 1991): 335; and Aaron Blair, Mustafa Dosemeci, and Ellen F. Heineman, "Cancer and Other Causes of Death Among Male and Female Farmers From Twenty-Three States," *American Journal of Industrial Medicine* 23 (1993): 729.

13. American Lung Association et al., *Indoor Air Pollution: An Introduction for Health Professionals* (Washington: EPA, 1994); and Leslie Dreyfous, "High-Tech Low: Are We Keeping Up With Our Brave New World?" Associated Press, 20 January 1991.

14. A. Sonia Buist and William M. Vollmer, "Reflections on the Rise in Asthma Morbidity and Mortality," *Journal of the American Medical Association* 264, no. 13 (13 October 1990): 1719–20; and Peter J. Gergen, Daniel I. Mullally, and Richard Evans III, "National Survey of Prevalence of Asthma Among Children in the United States, 1976 to 1980," *Pediatrics* 81, no. 1 (January 1988): 1–7.

15. Robert Repetto and Sanjay Baligan, *Pesticides and the Immune System: The Public Health Risks* (Washington: World Resources Institute, 1996), 18–56 passim.

16. Steven V. Arnold, "Synergistic Activation of Estrogen Receptor With Combinations of Environmental Chemicals," *Science* 272 (7 June 1996): 1489–91.

17. Inform, Inc., *Toxics Watch 1995,* (New York: Inform, 1995), 4.

18. George Peaff, "Dow Replaces DuPont to Lead Top 100 U.S. Chemical Producers," *Chemical & Engineering News,* 6 May 1996, 15–20.

19. James R. Davis, Ross C. Brownson, and Richard Garcia, "Family Pesticide Use in the Home, Garden, Orchard, and Yard," *Archives of Environmental Contamination and Toxicology* 22 (1992): 260.

20. Statement of Peter F. Guerrero, Associate Director for Environmental Protection Issues of the Resources, Community, and Economic Development Division, GAO, before the House Committee on Government Operations, Subcommittee on Environment, Energy, and Natural Resources, as reprinted in *Pesticides: 30*

Years Since Silent Spring—Many Long-Standing Concerns Remain (Washington: GAO, 1992), 2.

21. Jan Hollingsworth, "Covenant of Silence," *Tampa Tribune*, 18 December 1995, 1.

22. Jan Hollingsworth, "Judge Takes Benlate to Task," *Tampa Tribune*, 22 June 1996, 1.

23. Center for Responsive Politics, National Library on Money and Politics, from Federal Election Commission records.

24. Georgia-Pacific Corporation, pamphlet, n.d.

Chapter 1: Presumed Innocent

1. Christopher J. Bosso, *Pesticides & Politics* (Pittsburgh: University of Pittsburgh Press, 1987), 30.

2. On March 7, 1996, Ciba-Geigy and Sandoz Corporation announced that they intend to merge and form a new company called Novartis.

3. Committee on Comparative Toxicity of Naturally Occurring Carcinogens et al., *Carcinogens and Anticarcinogens in the Human Diet: A Comparison of Naturally Occurring and Synthetic Substances* (Washington: National Academy Press, 1996), iv, 289.

4. Theo Colburn, Dianne Dumanoski, and John Peterson Myers, *Our Stolen Future* (New York: Penguin Books USA, 1996).

5. Inform, Inc., *Toxics Watch 1995* (New York: Inform, 1995), 73–75.

6. Ibid., 70.

7. Susan Banks, "Hidden Dangers Make Pesticides Unhealthy Choice," *Pittsburgh Post-Gazette*, 4 June 1994.

8. David Grogan, "A Golfer's Mysterious Death," *People*, 23 June 1986.

9. Claudia Feldman, "Pretty...and Toxic?" *Houston Post*, 23 July 1995, 1.

10. Associated Press, "Chemicals Suspected in Illness," 17 April 1996.

11. Robert N. Proctor, *Cancer Wars: How Politics Shapes What We Know and Don't Know About Cancer* (New York: Basic Books, 1995), 27.

12. David Pimentel et al., "Environmental and Economic Costs of Pesticide Use," *Bioscience* 47, no. 10 (November 1992): 750.

13. Statement of Peter F. Guerrero, Director for Environmental Protection Issues of the Resources, Community, and Economic Development Division, GAO, before the Senate Committee on Appropriations, Subcommittee on VA, HUD, and Independent Agencies, "EPA's problems With Collection and Management of Scientific Data and Its Efforts to Address Them" (Washington: General Accounting Office, 12 May 1995), 4.

14. National Research Council, *Pesticides in the Diets of Infants and Children* (Washington: National Academy Press, 1993), 360–62.

15. Federal Insecticide, Fungicide, and Rodenticide Act as Amended, sec. 2 (bb).

16. Corrosion Proof Fittings v. EPA, 947 F.2d 1201 (5th Cir. 1991), 1201.

17. Ibid., 1201–30.

18. J. R. Condray, director of regulatory management, Monsanto, letter to EPA TSCA Section 8 (e) Coordinator, 15 July 1992.

19. Peter Voytek, vice president, Halogenated Solvents Industry Alliance, letter to EPA, TSCA Section 8 (e) Coordinator, 8 December 1992.

Chapter 2: Four Chemicals

1. An excellent source on the history of atrazine and Ciba-Geigy is the decision in Ciba-Geigy Corporation v. Internal Revenue Service, 85 T.C. 172, Docket No. 2381-78, United States Tax Court, 1985.

2. Bob Uhler, "Atrazine," *Journal of Pesticide Reform* 11, no. 4 (Winter 1991); n.p.

3. Ciba-Geigy Corporation v. Internal Revenue Service, 209.

4. EPA, "Pesticide Industry Sales and Usage: 1994 and 1995 Market Estimates," unreleased draft document dated March 1996.

5. Stewart Reeve, "Ciba Fields Counterattack on Atrazine's Special Review," *Farm Futures Magazine* 23, no. 1 (January 1995): 40.

6. EPA atrazine public docket, 25 April 1985.

7. EPA, Office of Pesticides, Hazard Evaluation Division, "Tox Oneliners," for atrazine studies, 1992, 1.

8. EPA atrazine public docket, 13 January 1987.

9. IARC Working Group on the Evaluation of Carcinogenic Risks to Humans, "Occupational Exposures in Insecticide Application, and Some Pesticides: Atrazine," IARC Monographs on the Evaluation of Carcinogenic Risks to Humans, vol. 53 (Lyon, France: World Health Organization, 1991): 449.

10. Ibid., 448–50.

11. H. Leon Bradlow et al., "Effects of Pesticides on the Ratio of 16 alpha/2-Hydroxyestrone: A Biologic Marker of Breast Cancer Risk," *Environmental Health Perspectives* 103, supp. 7 (October 1995): 147–50.

12. EPA atrazine public docket, 11 January 1988.

13. EPA atrazine public docket, 22 November 1988.

14. Paul Keck, *Missouri River Public Water Supplies Association 1991 Missouri River Monitoring Study*, unpublished study, 10.

15. Laurie Moyer and Joel Cross, *Pesticide Monitoring: Illinois EPA's Summary of Results 1985–1989*, (n.p.: Illinois EPA, 1990), 17.

16. R. D. Rowden et al., *Groundwater Monitoring in the Big Spring Basin 1992–1993: A Summary Review* (n.p.: Iowa Department of Natural Resources, 1995), 34; and R. D. Rowden et al., *Surface Water Monitoring in the Big Spring Basin 1986–1992: A Summary Review* (n.p.: Iowa Department of Natural Resources, 1995), 72–73.

17. D. A. Goolsby et al., "Occurrence of Herbicides and Metabolites in Surface Water, Ground Water, and Rainwater in the Midwestern United States," Proceedings, 1995 Annual Conference: Water Research, American Water Works Association, 18–22 June 1995, 588.

18. *U.S. Water News*, February 1991.

19. EPA atrazine public docket, 6 May 1992.

20. EPA atrazine public docket, 9 November 1994.

21. J. Alan Roberson, "Drinking Water Compliance Problems From Pesticides and Herbicides," American Water Works Association, 1994, 9.

22. Statement of Peter F. Guerrero, 2.

23. Marjorie Sun, "EPA Proposal on Alachlor Nears," *Science*, 12 September 1986, 1143.

24. William Marcus, "Alachlor: Health Advisory," *Toxicology and Industrial Health* 3, no. 3 (1987): 389–90.

25. EPA, "Alachlor Position Document 1: Initiation of Special Review," January 1985, 68.

26. "Alachlor Position Document 1," 26.

27. *Pesticide Monitoring*, 17.

28. EPA alachlor public docket, 15 August 1986.

29. Ibid.; and EPA, "Pesticide Fact Sheet for Alachlor," 14 December 1987, 3.

30. EPA, "National Pesticide Survey: Alachlor," Fall 1990, 2.

31. Monsanto Company, "Straight Talk About Lasso: An Environmental Report," 1991, 5.

32. K. M. Thompson, "Risk Assessment and the Dry-Cleaning Industry," in *The Greening of Industry: A Risk Management Approach*, ed. J. D. Graham, submitted for publication (1996), 3–4.

33. Ibid., 4, 39.

34. Transcript of EPA Dry Cleaning Public Meeting, 3 November 1993, 32–33.

35. Judith S. Schreiber et al., "An Investigation of Indoor Air Contamination in Residences Above Dry Cleaners," *Risk Analysis* 13, no. 3 (1993): 335–44.

36. Wendy Cohen, "Investigations of Groundwater Contamination by Perchloroethylene in California's Central Valley," *Proceedings of an EPA Design for the Environment International Roundtable: Pollution Prevention and Control in the Dry Cleaning Industry*, EPA, 27–28 May 1992, 63.

37. Judith S. Schreiber, "Predicted Infant Exposure to Tetrachloroethylene in Human Breastmilk," *Risk Analysis* 13, no. 5 (1993): 515–24.

38. Avima M. Ruder, Elizabeth M. Ward, and David P. Brown, "Cancer Mortality in Female and Male Dry-Cleaning Workers," *Journal of Occupational Medicine* 36, no. 8 (August 1994): 867.

39. Ann Aschengrau et al., "Cancer Risk and Tetrachloroethylene-Contaminated Drinking Water in Massachusetts," *Archives of Environmental Health* 48, no. 5 (September/October 1993): 284–92; and transcript of EPA Dry Cleaning Public Meeting, 3 November 1993, 128.

40. Patrick Kennedy, "Risk to Residents Living in Units Attached to Dry Cleaners," EPA Risk Management Memorandum, 11 May 1992, 1; and "Risk Assessment and the Dry Cleaning Industry," 1.

41. EPA, "Multiprocess Wet Cleaning: Cost and Performance Comparison of Conventional Dry Cleaning and and Alternative Process," September 1993, 1-1.

42. EPA, "Draft Brochure on Wet Cleaning Facilities," 26 March 1996.

43. "Multiprocess Wet Cleaning," 2-15.

44. "Risk to Residents," 1.

45. "Risk Assessment and the Dry-Cleaning Industry," 30.

46. "Chemical Profile: Formaldehyde," *Chemical Marketing Reporter*, 11 September 1995, 40–41.

47. Beat Meyer, *Urea-Formaldehyde Resins* (Reading, Mass.: Addison-Wesley, 1979), 4.

48. Ibid., 110, 185.
49. Ibid., 19, 196.
50. Harold Scarlett, "Formaldehyde Improvement Efforts Urged," *Houston Post*, 22 November 1981, sec. B, p. 7.
51. William D. Kerns et al., "Carcinogenicity of Formaldehyde in Rats and Mice After Long-Term Inhalation Exposure," *Cancer Research* 43 (September 1983): 4382; and James A. Swenberg et al., "Induction of Squamous Cell Carcinomas of the Rat Nasal Cavity by Inhalation Exposure to Formaldehye Vapor," *Cancer Research* 40 (September 1980): 3398–402.
52. Aaron Blair et al., "Epidemiologic Evidence on the Relationship Between Formaldehyde Exposure and Cancer," *Scandinavian Journal of Work, Environment, and Health* 16 (1990): 385–87.

Chapter 3: Science For Sale

1. Stewart Reeve, "Ciba Fields Counterattack on Atrazine's Special Review," *Farm Futures Magazine* 23, no. 1 (January 1995): 40.
2. Sanctions ruling of U.S. District Court Judge J. Robert Elliott, Case no. 4:95-cv-36 (JRE), The Bush Ranch Inc., et al. v. E. I. du Pont de Nemours and Company, 21 August 1995, 40.
3. Georgia-Pacific documents, 9 September 1980.
4. Internal memorandum summarizing findings of EPA/FDA inspection of IBT labs; and "Memorandum of Telecon Between Walter Hansen, S.I.S., HED-108 & David R. Foltz, Supervisory Investigator, Springfield Resident Post," 14 March 1978, 22.
5. "Memorandum of Telecon," 21.
6. Kenneth A. Konz, Assistant Inspector General for Audit, Office of the Inspector General, EPA, memorandum to Linda J. Fisher, Assistant Administrator for Prevention, Pesticides, and Toxic Substances, EPA, 30 September 1992.
7. Sanctions ruling of U.S. District Court Judge J. Robert Elliott, 40.
8. EPA alachlor public docket, 7 August 1985.
9. EPA alachlor public docket, 26 September 1985.
10. EPA alachlor public docket, 15 April 1985.
11. EPA alachlor public docket, 29 May 1985.
12. EPA alachlor public docket, 18 July 1985.
13. Georgia-Pacific documents, 27 March 1980.
14. Georgia-Pacific documents, 9 September 1980.
15. EPA formaldehyde documents, 19 October 1979.
16. FI documents, 7 February 1980.
17. FI documents, 4 March 1980.
18. FI documents, 7 January 1980.

19. Mary Ann Farrell, "Industry Awaiting Test Results," *Houston Evening Times*, 14 February 1980, sec. B, p. 14.
20. FI documents, 14 February 1979.
21. FI press release, "No Chronic Effects From Formaldehyde in Home, New

Study Reports," 26 February 1980; FI documents, 25 July 1979; and "Inhalation Study Asserts Formaldehyde Is Safe," *Chemical Engineering*, 3 May 1982, 23.

22. "No Chronic Effects From Formaldehyde in Home."

23. FI documents, 4 March 1980.

24. FI documents, 15 July 1980.

25. G. M. Rusch et al., "A 26-Week Inhalation Toxicity Study With Formaldehyde in the Monkey, Rat, and Hamster," *Toxicology and Applied Pharmacology* 68, no. 3 (May 1983): 329–43.

26. FI documents, 22 March 1979.

27. National Cancer Institute, "Cancer Facts: Formaldehyde," June 1994.

28. Irving J. Selikoff, Mt. Sinai Medical Center, letter to Aaron Blair, National Cancer Institute, 8 May 1980.

29. Philip J. Landrigan, Mt. Sinai Medical Center, letter to Eric Frumin, director of occupational safety and health, Amalgamated Clothing and Textile Workers Union, 24 February 1986.

30. James Gibson, ed., *Formaldehyde Toxicity*, Chemical Industry Institute of Toxicology Series (Washington: Hemisphere, 1983), xviii.

31. FI documents, 24 October 1979.

32. Rory B. Conolly et al., "Multidisciplinary, Iterative Examination of the Mechanism of Formaldehyde Carcinogenicity: The Basis for Better Risk Assessment," *CIIT Activities* 15, no. 12 (December 1995): 1; and "A 15-Year Index to *CIIT Activities:* Vols. 1–15 (1981–1995)," *CIIT Activities* 15, Index Supplement (December 1995): 5.

33. FI documents, 18 November 1981.

34. "Formaldehyde Largely Exonerated in Formaldehyde Institute Symposium," *Pesticide & Toxic Chemical News*, 10 November 1982, 3.

35. EPA formaldehyde documents, 3 February 1983.

36. Nicholas A. Ashford, "A Framework for Examining the Effects of Industrial Funding on Academic Freedom and the Integrity of the University," *Science, Technology, & Human Values* 8, no. 2 (Spring 1983): 16–23.

37. The authors used the MEDLINE database, the online version of the Index Medicus, to analyze 346 articles on alachlor, atrazine, perchloroethylene, or formaldehyde that appeared in major biomedical journals from 1989 through 1994 and were indexed as of February 1995. That set was designed to be comprehensive, containing every article in the index concerning alachlor (57), atrazine (100), perchloroethylene (62), or the toxicity of formaldehyde (127). The search for formaldehyde articles was narrowed to those that concerned toxicity because the word "formaldehyde" appears in thousands of abstracts unrelated to study of its effects on health because of the chemical's use as a biomedical preservative. The more common current name "perchloroethylene" was used in the search instead of the chemical's older name, "tetrachloroethylene," although some studies may have been overlooked because of this choice. Sixteen studies dealt with both alachlor and atrazine and were analyzed separately for each chemical. Forty-five studies could not be analyzed because the full article or an English translation was not readily available. Of the remaining articles, the analysis concluded that four on alachlor, 24 on atrazine, 32 on formaldehyde, and 21 on perc were irrelevant to the question of the chemical's effects on health. These

included studies on how to remediate groundwater or waste sites contaminated with such chemicals and studies in which the chemicals were used or mentioned that did not focus on the toxicity of the chemicals themselves—for example, the common test of the effect of medical anesthetics through irritation of rats with formaldehyde. All of the remaining 209 studies were analyzed to determine whether their results that tended to be favorable or unfavorable to the continued use of the chemicals or were ambivalent or too difficult to characterize. Study sponsors were determined from the acknowledgments sections included in the studies themselves. In cases where no sponsor was acknowledged but the study was conducted at a research institute, that institute was deemed the sponsor. In cases where multiple sponsors were listed, but only one could be characterized as either industry or independent, it was deemed the sponsor. In a few cases, a study's sponsor was not explicitly stated but could be reasonably inferred. No sponsor could be determined for 48 of the studies (five on alachlor, nine on atrazine, 27 on formaldehyde, and seven on perchloroethylene), and these were excluded from the final analysis. The full list of studies, including how they were classified by the authors, is accessible through the Center for Public Integrity's site on the World Wide Web (http://www.essential.org/cpi).

38. Orvin C. Burnside, "Weed Science—The Step Child," *Weed Technology* 7, no. 2 (April–June 1993): 515–18.

39. Numbers from a database created by the Center for Public Integrity using Internal Revenue Service Form 990 reports for the charitable foundations or other not-for-profit organizations associated with the companies.

40. EPA formaldehyde documents, 16 February 1982.

41. American Cancer Society, "Cancer Risk Assessment: Learning to Live With Cancer Risk," *Research News* 4, no. 3 (December 1995): 1.

42. Chemical Industry Institute of Toxicology 1994 Annual Report, 17.

43. EPA alachlor public docket, March 1985.

44. University Extension, University of Missouri–Columbia, press release, "'Farmers are not the bad guys': Crop chemical faces ban; found in drinking water," 24 February 1995.

45. University of Missouri–Columbia press release, "Pesticides-in-Water Scare Is 'Unfounded' Researchers Say," 1995.

46. Nancy Mays, College of Agriculture, Food, and Natural Resources, University of Missouri–Columbia, memorandum to authors, n.d.

47. David B. Baker et al., "Some Characteristics of Pesticide Transport in Lake Erie Tributaries," *Proceedings of the 26th Conference on Great Lakes Research, May 23–27, 1983* (Summary of conference), 22.

48. R. Don Wauchope et al., "Pesticides in Surface Water and Ground Water," *CAST Issue Paper*, no. 2 (April 1994): 1, 6.

49. David B. Baker, letter to the reader in "A Review of the Science, Methods of Risk Communication and Policy Recommendations in Tap Water Blues: Herbicides

in Drinking Water," by David B. Baker, R. Peter Richards, and Kenneth N. Baker, Heidelberg College Water Resources Program, 1994, n.p.

50. R. P. Richards et al., "Atrazine Exposures Through Drinking Water: Exposure Assessments for Ohio, Illinois, and Iowa," *Environmental Science Technology* 29, no. 2 (1995): 406–12.

51. Water Quality Laboratory, Heidelberg College, 1994 Annual Report, 16.

52. FI documents, 12 August 1982.

53. FI documents, 15 January 1981.

54. Gulf South Insulation v. CPSC, 701 F.2d 1137 (1983), 1145.

55. FI documents, 25 July 1979; Gary M. Marsh, "Proportional Mortality Among Chemical Workers Exposed to Formaldehyde," prepared for the Monsanto Company, 15 May 1981, submitted to the Office of the Surgeon General; and J. Donald Miller, Assistant Surgeon General, Director, to William R. Caffey, Manager, Epidemiology, Monsanto Company, 23 March 1982.

56. FI documents, 5 June 1979.

57. FI documents, 4 October 1978.

58. Numbers from a database created by the Center for Public Integrity using Internal Revenue Service Form 990 reports for the charitable foundations or other not-for-profit organizations associated with the companies.

59. IARC Working Group on the Evaluation of Carcinogenic Risks to Humans, "Occupational Exposures in Insecticide Application, and Some Pesticides: Atrazine," IARC Monographs on the Evaluation of Carcinogenic Risks to Humans, vol. 53 (Lyon, France: World Health Organization, 1991): 451.

60. A. Pinter et al., "Long-Term Carcinogenicity Bioassay of the Herbicide Atrazine in F344 Rats," *Neoplasma* 37, no. 5 (1990): 533–44.

61. "Occupational Exposures in Insecticide Application," 460.

62. EPA atrazine public docket, n.d.

63. Avima M. Ruder, Elizabeth M. Ward, and David P. Brown, "Cancer Mortality in Female and Male Dry-Cleaning Workers," *Journal of Occupational Medicine* 36, no. 8 (August 1994), 867.

64. Transcript of EPA Dry Cleaning Public Meeting, 3 November 1993.

65. EPA perc documents, 20 November 1993.

66. EPA perc documents, 19 November 1993.

67. EPA, "Environmental Equity: Reducing Risk for All Communities," vol. 2, *Supporting Documents*, Office of Policy Planning and Evaluation, June 1992, 1.

68. Ann Aschengrau et al., "Cancer Risk and Tetrachloroethylene-Contaminated Drinking Water in Massachusetts," *Archives of Environmental Health* 48, no. 5 (September/October 1993): 284–92; and transcript of EPA Dry Cleaning Public Meeting, 3 November 1993, 128.

69. *Cancer Wars*, 117n.

70. George W. Lucier, "Mechanism-Based Toxicology in Cancer Risk Assessment: Implications for Research, Regulation, and Legislation," *Environmental Health Perspectives* 104, no. 1 (January 1996): 84; NTP, "Workshop on Mechanism-Based Toxicology in Cancer Risk Assessment: Implications for Research, Regulation, and Legislation," draft of unpublished paper, 27 June 1995, 1; House Committee on

Appropriations, "Departments of Labor, Health and Human Services, and Education, and Related Agencies Appropriation Bill, 1995," report submitted by Neal Smith, 103d Cong., 2d sess., 1994, 57; and House Committee on Appropriations, Subcommittee on the Departments of Labor, Health and Human Services, Education and Related Agencies, transcript of hearings, 102d Cong., 2d sess., 1992, 20, 22, 27.

71. House Committee on Appropriations, "Departments of Labor, Health and Human Services, and Education, and Related Agencies Appropriation Bill, 1995," report submitted by Neal Smith, 103d Cong., 2d sess., 1994, 57; and House Committee on Appropriations, "Departments of Labor, Health and Human Services, Education, and Related Agencies Appropriations Bill, 1994," report submitted by William Natcher, 103d Cong., 1st sess., 1993, 63.

72. Videotape of the Workshop on Mechanism-Based Toxicology in Cancer Risk Assessment: Implications for Research, Regulation, and Legislation, NTP, Research Triangle Park, North Carolina, 11 January 1995.

73. Ibid.

74. Transcript of EPA Dry Cleaning Public Meeting, 3 November 1993, 174.

75. IARC, "Dry Cleaning, Some Chlorinated Solvents and Other Industrial Chemicals," *IARC Monographs* 63 (1995).

76. Ibid.

77. EPA formaldehyde documents, 30 June 1994.

78. EPA formaldehyde documents, 15 September 1992; and "Move Praised by Labor Union: SAB Panel Rejects Cuts in Formaldehyde Risks Despite EPA Office's Support," *Inside EPA*, 26 July 1991, 5.

79. Joan L. Gallagher, manager of TCSA compliance and international registrations, American Cyanamid Company, correspondence to EPA TCSA Section 8(e) Coordinator, 3 May 1991.

80. Cheryl Siegel Scott, epidemiologist, Carcinogen Assessment Statistics and Epidemiology Branch, Human Health Assessment Group, EPA, memo to Vanessa Vu, chief of oncology branch, Health and Environmental Assessment Division, EPA, n.d.

81. Mark H. Christman, counsel, E.I. Du Pont de Nemours & Co., submission to EPA TSCA Section 8 (e) Coordinator, 2 September 1992, and attachments.

82. Conolly et al., "Multidisciplinary, Iterative Examination of the Mechanism of Formaldehyde Carcinogenicity: The Basis for Better Risk Assessment," 7; and Dragana A. Andjelkovich, Derek B. Janszen, Rory B. Conolly, and Frederick J. Miller, "Formaldehyde Exposure Not Associated With Cancer of the Respiratory Tract in Iron Foundry Workers," *CIIT Activities* 15, no. 7 (July 1995).

83. Transcript of EPA Public Meeting on Formaldehyde Testing in New Homes, 28 January 1993, 63.

84. O. Hernandez et al., "Risk Assessment of Formaldehyde," *Journal of Hazardous Materials* 39 (1994): 172.

85. IARC, "Wood Dusts and Formaldehyde," *IARC Monographs* 62 (1995).

86. EPA atrazine public docket, n.d.

87. EPA atrazine public docket, 17 May 1985.

88. EPA alachlor public docket, 14 March 1996.

Chapter 4: Keeping the Watchdogs on a Short Leash

1. Transcript of EPA Public Meeting on Formaldehyde Testing in New Homes, 28 January 1993, 249–50.

2. Statement of Peter F. Guerrero, Associate Director for Environmental Protection Issues of the Resources, Community, and Economic Development Division, GAO, before the House Committee on Government Operations, Subcommittee on Environment, Energy, and Natural Resources, as reprinted in *Pesticides: 30 Years Since Silent Spring—Many Long-Standing Concerns Remain* (Washington: GAO, 1992), 1, 4.

3. EPA press release, "EPA Announces Rule Significantly Reducing Smog and Cancer-Causing Toxics From Landfills," 1 March 1996; EPA, "Design for the Environment, Dry Cleaning Project: Cleaner Clothes, Cleaner Neighborhoods, Cleaner Solutions," December 1994; "EPA Waste Chief Objects to Senate 'Micromanagement' of Water Regulations," *Bureau of National Affairs Environmental Law Update*, 11 September 1992, n.p.; and "EPA Says Responses Favorable on Disinfection By-Product Rule," *Bureau of National Affairs Daily Report for Executives*, 28 August 1992, n.p.

4. EPA alachlor public docket, 24 May 1984.

5. EPA alachlor public docket, 13 June 1984.

6. Corrosion Proof Fittings v. EPA, 947 F. 2d 1201 (5th Cir. 1991), 1201–30.

7. James Ramey, Chairman, FI, to John Hernandez, Deputy Administrator, EPA, 13 August 1981.

8. John Galloway and Barry Harger, memorandum to Representative Toby Moffett, Chairman, House Committee on Government Operations, Subcommittee on Environment, Energy, and Natural Resources, 1 October 1981; Moffett, memorandum to members of the Subcommittee, 6 October 1981; testimony of Dr. John Hernandez Jr., Deputy Administrator, EPA, before the House Committee on Government Operation, Subcommittee on Environment, Energy, and Natural Resources, 21 October 1981; FI documents, 30 September 1982 and 4 November 1982; "Chronology of EPA Controversy," United Press International, 25 March 1983.

9. Francis X. Clines and Bernard Weintraub, "Briefing," *New York Times*, 8 October 1981, sec. A, p. 22; Joanne Omang, "Panel Questions EPA Official Over Industry Meetings," *Washington Post*, 22 October 1981, sec. A, p. 1; and Howard Kurtz, "House Panel Probing Chemical Industry Influence on Todhunter," *Washington Post*, 12 March 1983, sec. A, p. 2.

10. Eliot Marshall, "EPA Indicts Formaldehyde, Seven Years Later," *Science* 236 (24 April 1987): 381.

11. See chapter 3.

12. *"Move Praised by Labor Union:* SAB Panel Rejects Cuts in Formaldehyde Risks Despite EPA Office's Support," *Inside EPA*, 26 July 1991, 5; transcript of EPA Public Meeting on Formaldehyde Testing in New Homes, 28 January 1993, 62–64; and EPA formaldehyde documents, 30 June 1994.

13. Nicholas R. Iammartino, vice president of public affairs, Borden, Inc., correspondence to Marianne Lavelle, 10 July 1996.

14. National Particleboard Association press release, "EPA/NPB 'Home Study'

Results Announced: Final Report Cites Low Formaldehyde Levels," 11 April 1996.

15. FI documents, 30 September 1980 and 11 February 1981.

16. Beat Meyer, *Urea-Formaldehyde Resins* (Reading, Mass.: Addison-Wesley, 1979), 243.

17. FI documents, 30 October 1980, 3.

18. CPSC, "An Update on Formaldehyde," October 1990, 11.

19. CPSC, National Injury Information Clearinghouse, 1992 to present.

20. Nicholas R. Iammartino, correspondence to M. Lavelle, 10 July 1996.

21. CPSC, National Injury Information Clearinghouse estimates report, Wood Paneling and Particle Board, 1994. Based on the number of actual patients treated in 1994 in a statistically representative sample of U.S. hospitals, the CPSC estimated that 90 persons across the nation would have gone to emergency rooms that year with complaints of "anoxia," a lack of oxygen because of exposure to wood paneling or particleboard. About 35 would have been treated for dermatitis or conjunctivitis complaints, according to the CPSC's estimate. Of the wood-product and particleboard injuries in the CPSC's database, the agency estimates that 115 were in mobile homes. This number, however, may include blunt-body injuries not related to formaldehyde. The CPSC warns that "extra caution" should be used in drawing conclusions from estimates drawn from statistics as small as those in its particleboard and wood-products estimates. "The Commission recognizes that estimates based on small samples—if carefully interpreted — are useful," it says in its explanation of its estimates.

22. Cases compiled from "The Exchange," a database maintained by the Association of Trial Lawyers of America in Washington, D.C.

23. EPA, OTS, "Formaldehyde Risk Assessment Update, Final Draft," 11 June 1991, 88.

24. EPA formaldehyde documents, 12 July 1991.

25. EPA formaldehyde documents, 18 March 1993.

26. Transcript of EPA Public Meeting on Formaldehyde Testing in New Homes, 28 January 1993, 35.

27. Ibid., 56.

28. EPA formaldehyde documents, 6 August 1993, 7 September 1993, and 1 December 1993.

29. National Particleboard Association press release, "Particleboard/MDF Industry and U.S. EPA Agree to Formaldehyde Pilot Home Study," 6 December 1993.

30. Glenn Hess, "Tests Narrowed on Formaldehyde," *Chemical Marketing Reporter*, 14 August 1995, 20.

31. "EPA/NPB 'Home Study' Results Announced: Final Report Cites Low Formaldehyde Levels."

32. Judy Schreiber, "Investigations of Indoor Air Contamination in Residences Above Dry-Cleaning Establishments," *Proceedings of an EPA Design for the Environment International Roundtable: Pollution Prevention and Control in the Dry Cleaning Industry*, EPA, 27–28 May 1992, 56.

33. EPA perc documents, 11 May 1992 and n.d.

34. EPA perc documents, 21 April 1994.

35. "Design for the Environment, Dry Cleaning Project: Cleaner Clothes."

36. EPA perc documents, 24 May 1994.

37. EPA perc documents, 29 June 1994.

38. EPA perc documents, 31 August 1994.

39. EPA perc documents, 6 September 1994.

40. EPA perc documents, 13 September 1994, n.d., and 17 July 1995.

41. EPA perc documents, 17 July 1995.

42. EPA perc documents, 18 October 1994.

43. EPA perc documents, 9 December 1994.

44. EPA perc documents, 12 December 1994.

45. EPA Perc documents, 7 March 1995.

46. EPA perc documents, 16 February 1995.

47. EPA perc documents, 3 March 1995.

48. EPA perc documents, n.d.

49. FI documents, 7 February 1980, 4.

50. Ibid.

51. Gulf South Insulation v. CPSC, 701 F.2d 1137 (1983); FI documents, 7 April 1982; "Three Sue Over Urea-Formaldehyde Foam Ban," *Chemical Week*, 28 April 1982, 52; Marjorie Sun, "Formaldehyde Ban Is Overturned," *Science* 220 (13 May 1983): 699; and Nicholas A. Ashford, William C. Ryan, and Charles C. Caldart, "Law and Science Policy in Federal Regulation of Formaldehyde," *Science* 222 (25 November 1983): 894.

52. CPSC, 1993 Annual Report, 10, E-7.

53. EPA, "Alachlor Position Document 1: Initiation of Special Review," January 1985, 69.

54. EPA alachlor public docket, 8 April 1985.

55. "Preliminary Determination to Cancel Registrations of Alachlor Products Unless the Terms and Conditions Are Modified; Availability of Technical Support Document and Draft Intent to Cancel," *Federal Register* 51, no. 195 (8 October 1986): 36169.

56. "III. Summary of Benefits Assessment and Agency Evaluation of Comments and Additional Data Received," *Federal Register* 52, no. 251 (31 December 1987): 49496.

57. Craig Osteen and Fred Kuch, *Potential Bans of Corn and Soybean Pesticides: Economic Implications for Farmers and Consumers*, Agricultural Economic Report no. 546, April 1986, 19.

58. J. Alan Roberson, Associate Director for Regulatory Affairs, Government Affairs Office, American Water Works Association, to Dan Barolo, Chief, Special Review Branch, Office of Pesticide Programs, EPA, 14 July 1992.

59. David Pimentel et al., "Environmental and Economic Impacts of Reducing U.S. Agricultural Pesticide Use," *Handbook of Pest Management in Agriculture, Vol. 1* (2d ed., 1991), 230–32, 244.

60. Transcript of EPA Scientific Advisory Panel meeting, 19 November 1986, 13–14.

61. Ibid., 39.

62. EPA alachlor public docket, 18 May 1987.

63. EPA alachlor public docket, 14 March 1996.

64. EPA Health Assessment Group, "Response to Issues and Data Groups on the Carcinogenicity of Tetrachloroethylene (Perchloroethylene)," Office of Health and Environmental Analysis, EPA, September 1991, 4.

65. EPA alachlor public docket, 27 June 1985.

66. C. L. Smith, Department of Agriculture, internal memorandum, 17 June 1985.

67. John H. Sullivan, Deputy Executive Director, American Water Works Association, to Carolyn Brickey, Senior Counsel, Senate Committee on Agriculture, Nutrition, and Forestry, 25 March 1992.

68. Department of Agriculture, "Atrazine in Surface Waters: A Report of the Atrazine Task Group to the Working Group on Water Quality," April 1992, 10–11.

69. The single exception is lead paint, a nearly ubiquitous presence in HUD's low-income housing.

70. FI documents, 7 January 1980, 6.

71. FI documents, 10 December 1982, 3.

72. FI documents, 14 November 1980, 5–6.

73. EPA perc documents, 2 November 1992 and 27 October 1992; and EPA, "National Emission Standards for Hazardous Air Pollutants for Source Categories: Perchloroethylene Dry Cleaning Facilities," *Federal Register* 58, no. 182 (22 September 1993): 49364–65; and EPA, "Economic Impact Analysis of Regulatory Controls in the Dry-Cleaning Industry: Final," September 1993, 2–4.

74. Using a list of trips taken by EPA employees compiled by the EPA's Office of General Counsel, the Center for Public Integrity created a database of trips sponsored by corporations, associations, and other groups with an interest in the four chemicals. Additionally, all trips sponsored by universities were included in this database.

Chapter 5: Making Friends in High Places

1. Gillian Sandford, *1988 CQ Almanac* (Washington: Congressional Quarterly Press, 1988), 140.

2. Margaret E. Kriz, "Pesticide Pressures," *National Journal*, 10 December 1988, 3126.

3. *1988 CQ Almanac*, 139.

4. Marianne Lavelle, "Gingrich Staffer, Moderates Clash Over Agency's Role," *National Law Journal*, 13 November 1995, sec. A, p. 6; and John E. Yang, "Regulatory Overhaul Put Off in House, 'Behind the Debate on the Environment,' GOP Postpones Action," *Washington Post*, 6 March 1996, sec. A, p. 4.

5. Center for Responsive Politics, National Library on Money and Politics, from Federal Election Commission records.

6. Numbers from a database created by the Center for Public Integrity using congressional financial disclosure reports, 1990–94.

7. Larry Makinson, "The Price of Admission: Campaign Spending in the 1994 Elections," Center for Responsive Politics, 1995, 1; and Philip M. Stern, *The Best Congress Money Can Buy* (New York: Pantheon Books, 1988), 9.

8. FI documents, 7 July 1980.

9. Barbara Coleman, *Through the Corridors of Power: A Guide to Federal Rulemaking* (Washington: OMB Watch, 1987) 31.

10. Center for Responsive Politics, National Library on Money and Politics, from Federal Election Commission records.

11. Numbers from a database created by the Center for Public Integrity using congressional financial disclosure reports, 1990–94.

12. Ibid.

13. Ibid.; and Center for Responsive Politics, National Library on Money and Politics, from Federal Election Commission records.

14. *1992 CQ Almanac* (Washington: Congressional Quarterly Press, 1992), 213.

15. Numbers from a database created by the Center for Public Integrity using congressional financial disclosure reports, 1990–94; and *1992 CQ Almanac*, 214.

16. Carol Browner, Administrator, EPA, to Representative John Dingell, 28 February 1995.

17. EPA perc documents, 13 October 1994.

18. EPA perc documents 28 February 1995, 4 April 1995, and 7 March 1995.

19. Center for Responsive Politics, National Library on Money and Politics, from Federal Election Commission records.

20. Federal Election Commission records.

21. Gov. Terry Branstad to Lee Thomas, Administrator, EPA, 12 March 1985.

22. Rep. Jim Leach to Lee Thomas, Administrator, EPA, 1 April 1985.

23. FI documents, 14 November 1980.

24. Statement of R. Josh Lanier, National Insulation Certification Institute, before the Senate Committee on Commerce, Science, and Transportation, Subcommittee on Consumer Issues, 1–7 April 1981, on the reauthorization of the CPSC.

25. FI documents, 30 June 1981.

26. FI documents, 25 June 1981.

27. EPA perc documents, 16 December 1993, 3.

28. EPA perc documents, 20 November 1993, 14.

29. Federal Insecticide, Fungicide, and Rodenticide Act as Amended, sec. 2 (bb).

30. Mary Cohn, ed., *1984 CQ Almanac* (Washington: Congressional Quarterly Press, 1984), 365; and Christopher J. Bosso, *Pesticides & Politics* (Pittsburgh: University of Pittsburgh Press, 1987), 227.

31. EPA, "Fact Sheet on February 8, 1995, Court Approved Settlement Agreement on Pesticides Affected by the Delaney Clause," 10 February 1995.

32. Kelsey Wirth and Frank Schima, *The Pesticide PACs: Campaign Contributions and Pesticide Policy* (Washington: Environmental Working Group, 1994), 2.

33. Corrosion Proof Fittings v. EPA, 947 F. 2d 1201 (5th Cir. 1991), 1215–17.

34. GAO, "Toxic Substances Control Act: Legislative Changes Could Make the Act More Effective," September 1994, 21.

35. FI documents, 14 November 1980.

Chapter 6: Justice Denied

1. Cynthia Hubert, "Reports on Water Stir Few Ripples at Rathbun," *Des Moines Register*, 28 July 1991, 1.

2. R. Munger et al., "Birth Defects and Pesticide-Contaminated Water Supplies in Iowa," presented at the 25th Annual Meeting of the Society of Epidemiologic Research in Minneapolis, Minnesota, 1992.

3. Wendy Cohen, "Investigations of Groundwater Contamination by Perchloroethylene in California's Central Valley," *Proceedings of an EPA Design for the Environment International Roundtable: Pollution Prevention and Control in the Dry Cleaning Industry*, EPA, 27–28 May 1992, 63–65.

4. Transcript of EPA Dry Cleaning Public Meeting, 3 November 1993, 30.

5. Inform, Inc., *Toxics Watch 1995* (New York: Inform, 1995), 37–44.

6. EPA perc documents, 18 November 1993.

7. Gregory W. Diachenko, "Perchloroethylene Levels in Foods Obtained Near Drycleaning Establishments," *Proceedings of an EPA Design for the Environment International Roundtable: Pollution Prevention and Control in the Dry Cleaning Industry*, EPA, 27–28 May 1992, 49–50.

8. American Lung Association et al., "Indoor Air Pollution: An Introduction for Health Professionals," EPA, n.d., 14.

9. Richard Wiles et al., *Tap Water Blues: Herbicides in Drinking Water*, Environmental Working Group and Physicians for Social Responsibility, 1994, 64.

10. Ibid., 1.

11. Marjorie Sun, "EPA Proposal on Alachlor Nears," *Science* 233 (12 September 1986): 1143.

12. National Cancer Advisory Board, "Cancer at a Crossroads: A Report to Congress for the Nation," September 1994, 5–6; and Associated Press, "Cancer War Has Stalled, a Panel Says," *The New York Times*, 30 September 1994, A11.

13. U.S. Judicial Panel on Multidistrict Litigation, Docket 1042: In re Carbonless Paper Products Liability Litigation, Attorney Service List, 23 November 1994.

14. Cipollone v. Liggett Group Inc., 112 S.Ct. 2608 (1992).

15. Certificate of Dissolution of Formaldehyde Institute, Inc., Under Section 1003 of the Not-for-Profit Corporation Law, 20 December 1993.

16. George Snyder, "Yuroks Fear Cancer From Spraying," *San Francisco Examiner*, 7 July 1992, 1.

17. David Ranii, "Formaldehyde Suits Erupt," *National Law Journal* 4, no. 35 (10 May 1982): 1; and Lippes interview.

18. FI documents, 7 April 1982.

19. Graham v. Canadian National Railway Company, 749 F.Supp. 1300, 5 November 1990.

20. Department of Housing and Urban Development, "Manufactured Home Construction and Safety Standards: Final Rule," *Federal Register* 49 (9 August 1984): 31996.

21. EPA alachlor public docket, 20 March 1987.

22. Joseph P. Thomas, "Defending Against Formaldehyde Exposure Claims," *For the Defense*, February 1995, 31.

23. FI documents, 26 February 1981.

24. Transcript of EPA Dry Cleaning Public Meeting, 3 November 1993, 192, 198.

25. Testimony of Thomas F. Nicholson, Executive Vice President, West

Group, in Westfarm Associates v. International Fabricare Institute, 92-9, U.S. District Court, Maryland.

26. IFI documents, 2 April 1991.

27. IFI documents, 12 February 1991.

28. IFI documents, 1988.

29. IFI documents, 5 May 1989.

30. Ivan J. Andrasik, deposition for Westfarm Associates v. International Fabricare Institute, 17 July 1992.

31. IFI documents, May 1990.

Chapter 7: The PR Juggernaut

1. John C. Stauber and Sheldon Rampton, *Toxic Sludge Is Good for You* (Monroe, Maine: Common Courage Press, 1995), 124.

2. David Bianco, *PR News Casebook* (Detroit: Gale Research, 1993), 440.

3. Richard Rabin, "Warnings Unheeded: A History of Child Lead Poisoning," *American Journal of Public Health* 79, no. 12, 1668; and Lead Industries Association advertisement in *Better Homes and Gardens*, March 1940, 72.

4. *PR News Casebook*, 1147.

5. *PR News Casebook*, 1148.

6. "Silver Anvil Award Winners," *PR Services*, June 1994, n.p.; and "Silver Anvil Programs Show Power of Public Relations," *Public Relations Journal*, June/July 1994, 8.

7. Jim Donahue, "Another Tree, Another Dollar: Rampant Expansion at Georgia-Pacific," *Multinational Monitor*, October 1990, 30.

8. Ed Rieke, Monsanto Agricultural Products Company, to "Grower," 18 February 1985.

9. FI pamphlet, "A Building Block of Our Society," n.d.

10. R. Bruce Walters, Chairman, Texas Mobile Home Association, to Vanply et al., 27 October 1977.

11. Minutes of the National Particleboard Association Board of Directors, 9 November 1977.

12. Georgia-Pacific documents, 23 February 1982.

13. Center for Emissions Control press release, "Re: Dry Cleaning With Perchloroethylene," 13 May 1994.

14. "Profiles of Top Environmental PR Firms," *O'Dwyer's PR Services Report 9*, no. 2 (February 1995): 41.

15. Richard E. Stuckey, "1994 Was Great; 1995 Will Be Better," *NewsCAST*, Winter, 1995, 4.

16. "Science, Benefits Should Stay Atrazine," *Farm Journal*, February 1995, 52.

17. "Recipes for the Best Broadcasters: Five NAFB Members Honored for Cooking Up Unique Ways to Serve Industry," *Agri Marketing*, March 1994, 36.

18. EPA alachlor public docket, 19 November 1985, 21 November 1985, 4 December 1985, and n.d.

19. EPA alachlor public docket, 14 September 1992.

20. FI documents, 23 October 1981 and 3 November 1982.

21. Ibid.

22. FI documents, 18 November 1991.

23. Bill Lichtenstein, producer, and Peter Lance, correspondent, "The Danger Within," *20/20*, Show #205, 4 February 1982.

24. FI documents, 20 January 1982.

25. FI documents, 17 February 1982.

26. FI documents, 11 February 1982.

27. FI documents, 17 March 1982.

28. David Ranii, "Formaldehyde Suits Erupt," *National Law Journal* 4, no. 35 (10 May 1982): 1.

29. Mark Golden, producer, and Peter Lance, correspondent, "Have You Ever Wondered?" *20/20*, Show #219, 20 May 1982.

30. Transcript of EPA Dry Cleaning Public Meeting, 4 November 1993, 9–10, 17–18.

31. "1996 North American Ag Communications Agencies," *Agri Marketing*, May 1996, 18.

32. Ibid., 42.

33. EPA atrazine public docket, 1995.

34. FTC perc documents, 28 September 1994.

35. FTC perc documents, 28 September 1994; and Proposed Care Labeling Rule, *Federal Register* 59, no. 218 (14 November 1994): 57241.

36. FTC perc documents, n.d.

37. FTC perc documents, 16 December 1971.

38. FTC perc documents,15 October 1976.

Chapter 8: Assessing the Alternatives

1. John Bower, *The Healthy House* (New York: Lyle Stuart, 1993), 151.

2. Michael D. Koontz et al., "Residential Indoor Air Formaldehyde Testing Program: Pilot Study—Final Report," Geomet Technologies, Inc., 21 March 1996, 6–12, 6–13.

3. National Particleboard Association press release, "EPA/NPB 'Home Study' Results Announced: Final Report Cites Low Formaldehyde Levels," 11 April 1996.

4. "Formaldehyde-Free Interior Grade MDF," *Environmental Building News* 1, no. 1.

5. Data tables summarizing the Practical Farmers of Iowa studies were provided by Rick Exner, agronomist, Iowa State Univiersity Cooperative Extension Service.

6. Steve Peters, Rhonda Janke, and Mark Bohlke, "Rodale's Farming Systems Trial 1986–1990," Rodale Institute Research Center, 1992, 1.

7. Edward G. Smith et al., "Impacts of Chemical Use Reduction on Crop Yields and Costs," Agricultural and Food Policy Center, Department of Agricultural Economics, Texas A&M University, 1989, 11.

8. EPA, "Multiprocess Wet Cleaning: Cost and Performance Comparison of Conventional Dry Cleaning and An Alternative Process," September 1993, ES-iv.

Chapter 9: Fixing the System

1. Ronald Begley, "Buyers' Green Demands Challenge Suppliers," *Chemical Week*, 232 August 1995, 43–44.

2. Chemical Specialties Manufacturers Association, "Household Products Smart Choice, Home Mixtures Potential Danger," press release, 2 October 1995.

3. Statement of Leslie A. Brueckner, Staff Attorney, Trial Lawyers for Public Justice, before the Advisory Committee on Civil Rights of the Judicial Conference of the United States, 9 February 1996.

4. "Comments of the Product Liability Advisory Council on Proposed Changes to Rule 29 and Rule 35 of the Federal Rules of Appellate Procedure," attached to letter from Hugh F. Young Jr., Executive Director, Product Liability Advisory Council, to Peter G. McCabe, Secretary, Committee on Rules of Practice and Procedure, Administrative Office of the U.S. Courts, 29 February 1996.

5. "TLPJ Wins Major Secrecy Battle," *Public Justice*, Spring 1995, 1; "Battle Over Federal Secrecy Rules Is Revived: Committee Again Proposes to Permit Secrecy Without 'Good Cause,'" *Public Justice*, Summer 1995, 7; "Rule Changes Threaten to Increase Court Secrecy: Weakening of 'Good Cause' Requirement Proposed Again," *Public Justice*, Fall 1995, 10.

6. Ibid.

7. Nan Aron, President, Alliance for Justice, to Peter G. McCabe, Secretary, Committee on Rules of Practice and Procedure, Administrative Office of the U.S. Courts, 28 February 1996.

8. Paul Orum and Alair MacLean, "Progress Report: Community Right-to-Know," OMB Watch, 30 July 1992, n.p.

9. "Successful Model," *Environmental Health Letter* 34, no. 17 (14 August 1995): 139; David Roe and Gilbert Omenn, "California Has Successful Model of Regulatory Risk Assessment," *Seattle Post Intelligencer*, 25 July 1995, sec. A, p. 9.

Chapter 10: The Cost of Toxic Deception

1. Jim Drinkard, "Going All Out to Stop Curbs on Chlorine in Water," Associated Press, April 23, 1996.

2. Environmental Research Foundation, "Bill Gaffey's Work," *Rachel's Environment & Health Weekly*, May 16, 1996.

3. David Lyons, *National Law Journal*, June 24, 1996. A6.

4. "Benlate, Bad Science, and Justice," *Tampa Tribune*, July 15, 1996.

5. Thomas Clavin and Ziba Kashef, "Danger on Our Doorstep," *McCall's*, August 1993, 95.

6. Michael G. Wagner, "Study Warns of Pesticide Risk to O.C. Schoolkids," *Los Angeles Times* 8 February 1996, A-1.

7. Peter G. Sparber letter to Methyl Bromide Working Group, 17 June 1995.

8. Robin Harvey, "Something Bugging You? Insect Repellent Containing the Pesticide Deet Can, In Some Cases, Cause Irritation and Other Neurological Problems," *Toronto Star*, 18 July 1996, B5.

Appendix: Four Chemicals —
A Resource Guide

Manufacturers, trade associations, federal agencies, and environmental organizations have all played key roles in the long battles over the regulation of alachlor, atrazine, formaldehyde, and perchloroethylene. Here are descriptions of some of the major players.

Alachlor

Monsanto Company (http://www.monsanto.com)
800 North Lindbergh Boulevard
St. Louis, Missouri 63167
(314) 694-1000
Monsanto is the nation's fourth-largest chemical company. Originally founded to manufacture saccharin, Monsanto also invented Astroturf and sells NutraSweet. In 1995, the company generated $8.9 billion in sales. Chemical and agricultural products accounted for 69 percent of Monsanto's sales in 1995 and for about 77 percent of its operating income.

Chairman, President, and Chief Executive Officer: Robert B. Shapiro

Monsanto Agricultural Company
President, Crop Protection: Arnold Donald
E-mail: awdona@monsanto.com

Consumer Affairs: (800) 332-3111

Company employees registered as Washington lobbyists:

Vice President, Government Affairs: Linda Fisher
Vice President, International Government Affairs: Toby Moffett

Director, Agricultural Affairs: Chester T. Dickerson Jr.
Director, Government Affairs: Thomas Helscher
Director, Regulatory Affairs: Patricia Kenworthy
Director, Agriculture Regulation: W. Martin Strauss

Washington Office
700 14th Street, N.W.
Suite 1100
Washington, D.C. 20005
(202) 783-2460

Lobbying, law, and PR firms retained by Monsanto, 1994–95:

Arter & Hadden (Dennis Eckart)
The Duberstein Group
Fleishman-Hillard, Inc. (Henry W. Hubbard, Jim Mulhern)
Foreman & Heidepriem (Carol T. Foreman)
The Grizzle Company
Hill and Knowlton
Hogan & Hartson (William House, Humberto Pena, Clayton Yeutter)
Stewart and Stewart (James Cannon, Jr., Eugene Stewart)
Winthrop, Stimson, Putnam & Roberts (Raymond Calamaro)

Atrazine

Ciba-Geigy Corporation (http://www.ciba.com)
540 White Plains Road
P.O. Box 2005
Tarrytown, New York 10591
(914) 785-2000
Ciba-Geigy Corporation is a wholly owned subsidiary of Ciba-Geigy Ltd.,
which is headquartered in Basle, Switzerland, and is one of the world's
largest drug and chemical companies. Ciba-Geigy specializes in health and
agricultural chemicals. Geigy scientist Paul Muller won the Nobel Prize in
1939 for inventing DDT. Its current products include Efidac, Maalox, and
various other chemicals. Ciba-Geigy Ltd. has operations in more than 80
countries worldwide, generating sales in excess of 22.6 billion Swiss francs.
Agricultural products accounted for about 21 percent of Ciba's worldwide
sales in 1994. Ciba-Geigy announced in 1996 that it is merging with Sandoz
to form a company called Novartis.

Chairman: Richard Barth
President and Chief Executive Officer: Douglas Watson

Consumer Affairs: Nancy Caspar
581 Main Street
Woodbridge, New Jersey 07095
(908) 602-6800

Company employees registered as Washington lobbyists:

Vice President, Government Relations: George Rolofson
Director, Public Affairs: Ralph Loomis
Manager, Federal Government Relations: Patrick Donnelly
Managers, Federal Legislative Affairs: Glenn S. Ruskin, David P. Drake, Margaret Lyons
Manager, Communications: Deborah B. Myers
Executive Director, Public Affairs and Communication: David Celmer

Washington Office
1747 Pennsylvania Avenue, N.W.
Suite 700
Washington, D.C. 20006
(202) 293-3019

Lobbying, law, and PR firms retained by Ciba-Geigy, 1994–95:

Sutherland, Asbill & Brennan
R. Duffy Wall and Associates

Formaldehyde

Borden, Inc./Borden Chemicals and Plastics, L.L.P., Inc.
180 East Broad Street
Columbus, Ohio 43215
(614) 225-4000
Borden has been one of the nation's leading dairy companies since it was founded in 1857. Borden is best known through its mascot, Elsie the Cow. Its other well-known products include Krazy Glue, Cracker Jack, and Eagle Brand. Borden generated in excess of $5.6 billion in sales in 1994. That same year, packaging and industrial products accounted for 36 percent of the company's sales.

Borden, Inc.
Chairman: Frank J. Tasco
President and Chief Executive Officer: C. Robert Kidder

Borden Chemicals and Plastics
Chairman, President, and Chief Executive Officer: Joseph M. Saggese

Consumer Affairs: Borden Consumer Response
(800) 426-7336

Jayne Wieger
Borden Chemicals Customer Service
155 West A Street
Building A2
Springfield, Oregon 97477
(503) 744-3201

Lobbying, law, and PR firms retained by Borden, 1994–95:

Dross, Levenstein, Perilman and Kopstein (Herbert Levenstein)
Hooper, Hooper, Owen & Gould (James C. Gould, Daryl H. Owen)
Keck, Mahin and Cate (Thomas Jolly)
McDermott, Will and Emery (Gregory F. Jenner, Ronald L. Platt)
Sidley & Austin (Gabriel Adler, Michael Nemeroff)

E. I. du Pont de Nemours and Company (http://www.dupont.com)
1007 Market Street
Wilmington, Delaware 19898
(302) 774-1000
E-mail:info@dupont.com
DuPont is the nation's number-one chemical company. The company's long
list of inventions includes neoprene synthetic rubber (1931), Lucite (1937),
nylon (1938), Teflon (1938), and Dacron; it also produces Lycra, Mylar, and
Tyvek. DuPont and its subsidiaries conduct operations in about 70 countries
worldwide and generated $39.3 billion in sales in 1994. Chemicals accounted
for about 10 percent of DuPont's worldwide sales in 1994 and for about 14
percent of the company's operating income.

Chairman and Chief Executive Officer: John Krol

Consumer Affairs: Dr. Evan Riehl
(302) 452-1206

Company employees registered as Washington lobbyists:

Senior Washington Representative: Trudy M. Bryan
Washington Representatives: Celeste Boykin, Rodney J. McAlister
Director, International Trade and Investment: Robert M. Heine
Vice President, Federal Affairs: Mark D. Nelson
Senior Washington Counsel: Ellan K. Wharton

Washington Office
1701 Pennsylvania Avenue, N.W.

Suite 900
Washington, D.C. 20006
(202) 728-3600

Lobbying, law, and PR firms retained by DuPont, 1994–95:

Burson-Marsteller
Downey Chandler, Inc. (Rod Chandler, Thomas J. Downey, Margaret Hayden, Stephen Cooper, John Peter Olinger)
Wilmer, Cutler and Pickering (John D. Greenwald)
Winston & Strawn (LaJauna Wilcher)
Winthrop, Stimson, Putnam & Roberts (G. Lawrence Atkins, Kristin Bass)

Hoechst Celanese Corporation (http://www.hcc.com)
Route 202-206
P.O. Box 2500
Somerville, New Jersey 08876
(908) 231-2000
(800) 235-2637
E-mail: info@bwmail1.hcc.com

Hoechst Celanese is the largest affiliate of Hoechst Group, the world's largest chemical manufacturer. Hoechst's products include animal-feed additives, herbicides, textiles, and chemicals for the electronics industry. In 1994, the Hoechst Group and its subsidiaries generated in excess of 49.6 billion deutschmarks. Chemicals accounted for 27 percent of the company's total sales, and for 25 percent of its operating profits.

Chairman: Gunter Metz
President and Chief Executive Officer: Karl G. Engels

Consumer/Public Affairs
Lewis Alpaugh
Hoechst Celanese Corporation
Route 202-206
P.O. Box 2500
Somerville, New Jersey 08876
(908) 231-2880

Company employees registered as Washington lobbyists:

Vice President, Government Relations: Donald R. Greeley
Manager, Congressional Affairs: Robert Carpenter
Manager, Regulatory Affairs: W. Anthony Shaw
Director, Government Market Development: Eugene Steadman Jr.

Washington Office

919 18th Street, N.W.
Suite 700
Washington, D.C. 20006
(202) 296-2890

Lobbying, law, and PR firms retained by Hoechst Celanese, 1994–95:

Akin, Gump, Strauss, Hauer and Feld
Caplin and Drysdale
Hunton and Williams
Koleda Childress & Company (James Childress)
McGuiness & Williams (William N. LaForge)
Wilmer, Cutler and Pickering

Georgia-Pacific Corporation
133 Peachtree Street, N.E.
Atlanta, Georgia 30303
(404) 652-4000
Georgia-Pacific, the nation's top manufacturer, distributor, and wholesaler of building products, owns more than six million acres of timberland in the United States and Canada. Georgia-Pacific produces a wide array of wood-related products, including doors, particleboard, copy paper, and bath tissues. The company generated $12.7 billion in sales in 1994. Building products accounted for about 60 percent of sales and 80 percent of operating income.

Chairman and Chief Executive Officer: A.D. Correll

Consumer Affairs: (800) 477-2737

Company employees registered as Washington lobbyists:

Director, Federal Regulatory Affairs: Patricia Hill
Senior Director, Government Affairs: David T. Modi
Senior Government Affairs Representative: Denise M. O'Donnell
Vice President, Government Affairs: John M. Turner
Director, Environmental Policy, Training, and Regulatory Affairs: Susan F. Vogt

Washington Office
Georgia-Pacific Corporation
1875 I Street, N.W.
Suite 775
Washington, D.C. 20006
(202) 659-3600

Lobbying, law, and PR firms retained by Georgia-Pacific, 1994–95:

Burson-Marsteller
Dewey Ballantine (John J. Salmon)
McNair and Sanford
Winthrop, Stimson, Putnam & Roberts (G. Lawrence Atkins, Kristin Bass)

Weyerhaeuser Company
Tacoma, Washington 98477
(206) 924-2345
Weyerhaeuser Company is the world's largest private owner of softwood timber. The company's products include doors, particleboard, newsprint, and linerboard. Weyerhaeuser generated $10.4 billion in sales in 1994. Timberlands and wood products accounted for 48 percent of sales in 1994 and 82 percent of operating income. Although Weyerhaeuser does not produce formaldehyde, it is a large user and has been heavily involved in lobbying over formaldehyde regulation and testing.

President and Chief Executive Officer: John W. Creighton Jr.

Consumer Affairs: Kate Tate
3363 Weyerhaeuser Way South
Federal Way, Washington 98003
(206) 924-5304

Company employees registered as Washington lobbyists:

Vice President, Federal Affairs: Frederick S. Benson III
Managers, Government Affairs: Heidi E. Biggs, Paul Schlegel

Washington Office
Weyerhaeuser Company
1100 Connecticut Avenue, N.W.
Suite 530
Washington, D.C. 20036
(202) 293-7222

Lobbying, law, and PR firms retained by Weyerhaeuser, 1994–95:

Van Scoyoc Associates (H. Stewart Van Scoyoc)

Perchloroethylene

PPG Industries, Inc.
One PPG Place
Pittsburgh, Pennsylvania 15272

(412) 434-3131
http://www.ppg.com
E-mail: webmaster@ppg.com
PPG is the nation's leading glass manufacturer. The company's products include flame retardants, sealants, optical resins, and various sorts of glass. PPG generated $6.3 billion in sales in 1994. Chemicals accounted for about 20 percent of total sales in 1994 and about 23 percent of operating income.

Chairman of the Board and Chief Executive Officer: Jerry E. Dempsey
President and Chief Operating Officer: Raymond W. LeBoeuf

Consumer Affairs: Ken Lee
(412) 434-2604

Lobbying, law, and PR firms retained by PPG, 1994–95:

Baker & Hostetler
Baker, Donelson, Bearman and Caldwell
Stewart and Stewart (Terrence P. Stewart)
R. Duffy Wall and Associates (R. Duffy Wall)

Vulcan Materials Company
One Metroplex Drive
Birmingham, Alabama 35209
(205) 877-3000
Vulcan Materials, the parent company of Vulcan Chemicals, is the leading commercial producer of crushed stone. Vulcan produces chlorinated hydrocarbons, chlorine, and crushed rocks. Vulcan and its subsidiaries generated about $1.3 billion in sales in 1994. Chemicals accounted for 33 percent of sales in 1994.

Vulcan Chemicals
P.O. Box 530390
Birmingham, Alabama 35253
(205) 877-3484

Chairman and Chief Executive Officer: Herbert A. Sklenar

Consumer/Public Affairs: David Donaldson
Vulcan Materials
P.O. Box 530390
Birmingham, Alabama 35253
(205) 877-3021

Company employees registered as Washington lobbyists:

Director, Government Relations: E. John Wilkinson

Washington Office
Vulcan Chemicals
1899 L Street, N.W.
Suite 500
Washington, D.C. 20036
(202) 293-0635

Lobbying, law, and PR firms retained by Vulcan, 1994–95:

Hogan and Hartson (Kenneth Farber, Michael, Gilliland, W. Michael House, Gordon Martin, Christine Warnke)
Mayer, Brown and Platt (John C. Berghoff Jr., Julian DíEsposito, Roger J. Kiley, Amy L. Nathan)
Patton, Boggs (George Schutzer, James Christian)
Law Office of Luther J. Strange III (Luther J. Strange III)

Dow Chemical Company (http://www.dow.com)
Dow Center
2030 Dow Center
Midland, Michigan 48674
(517) 636-1000
Dow, the nation's number-two chemical company, has more than 90 manufacturing plants in 30 countries. The company introduced Saran Wrap, its first major consumer product, in 1953, and is the nation's top producer of caustic soda. Its Dow Corning venture, which produced breast implants, prompted a multimillion-dollar, multiyear lawsuit. Dow and its subsidiaries generated $20.2 billion in sales in 1995. Chemicals accounted for about 37 percent of Dow's worldwide sales and for about 41 percent of the company's operating income.

Chairman of the Board and Chief Executive Officer: Frank P. Popoff
President and Chief Operating Officer: William S. Stavropoulos

Consumer Affairs: Customer Information Group
Dow Chemical Company
P.O. Box 1206
Midland, Michigan 48674
(800) 258-2436
(517) 832-1556

Company employees registered as Washington lobbyists:

Manager, Government Relations: Paul N. Cicio
Vice President, Environmental and Regulatory Issues: Wilma Delaney
Vice President, Trade and Commercial Issues: Frank Farfone

Tax Manager, Government Relations: Laura M. Floam
Managers, Government Relations, Environmental Policy: Margaret Rogers, John R. Ulrich
Directors, Federal Technology/Business Development: Larry Sams, Paul Stone
Public Affairs Manager: Vicky M. Suazo

Washington Office
1776 I Street, N.W.
Suite 575
Washington, D.C. 20006
(202) 429-3400

Lobbying, law, and PR firms retained by Dow, 1994–95:

Bergner, Bockorny, Clough & Brain (John M. Clough Jr.)
Bonner & Associates
The Direct Impact Company
The Jefferson Group, Inc. (Thomas R. Donnelly Jr.)
Ketchem Public Relations
Koleda Childress & Co. (James M. Childress)
WinCapitol, Inc.
Winthrop, Stimson, Putnam & Roberts (G. Lawrence Atkins, Kristin Bass)

Imperial Chemicals Industries, PLC (http://www.demon.co.uk/ici)
Millbank
London, England SW1P3JF
0171 834 4444
Imperial Chemicals Industries is the largest manufacturer of industrial explosives in the world. Its products also include polymers, polyurethane, and acrylics. ICI generated sales of about 9.1 billion in British pounds in 1994. Industrial chemicals accounted for about 41 percent of sales in 1994 and for about 43 percent of operating income.

Chairman: Ronald Hampel
Chief Executive Officer: Charles Miller Smith

Federal Regulators

Environmental Protection Agency (http://www.epa.gov)

Office of Prevention, Pesticides, and Toxics
(http://www.epa.gov/docs/pesttoxics.html)
401 M Street, S.W.
Washington, D.C. 20460
(202) 260-1023

Office of Pesticide Programs (http://www.epa.gov/docs/pesttoxics.html)
1921 Jefferson Davis Highway
Arlington, Virginia 20376
(703) 308-7022

Office of Air and Radiation (http://www.epa.gov/oar/oarhome.html)
401 M Street, S.W.
Washington, D.C. 20460
(202) 260-7400
E-mail: link.tom@epamail.epa.gov

Department of Agriculture (http://www.usda.gov)

Agricultural Research Service (http://www.ars-grin.gov:80/ars/ars.html)
14th Street and Independence Avenue, S.W.
Washington, D.C. 20250
(301) 344-2340

Cooperative State Research Service (http://www.reeusda.gov)
24th Street and Independence Avenue, S.W.
Washington, D.C. 20250
(202) 720-6133

Occupational Safety and Health Administration (http://www.osha.gov)
200 Constitution Avenue, N.W.
Washington, D.C. 20210
(202) 219-8151

Consumer Product Safety Commission (http://www.cpsc.gov)
4330 East West Highway
Bethesda, Maryland 20814
(301) 504-0213

Food and Drug Administration (http://www.fda.gov)
5600 Fishers Lane
Room 1685
Rockville, Maryland 20857
(301) 443-5006

Congressional Committees

Senate Agriculture Committee
(http://www.senate.gov/committee/agriculture.html)
Senate Russell Office Building
Washington, D.C. 20510
(202) 224-2035

Senate Committee on Environment and Public Works
Senate Dirksen Office Building
Washington, D.C. 20510
(202) 224-6176

Senate Committee on Labor and Human Resources
(http://www.senate.gov/committee/labor.html)
Senate Dirksen Office Building
Washington, D.C. 20510
(202) 224-6770

House Committee on Agriculture
Longworth House Office Building
Washington, D.C. 20515
(202) 225-2171

House Committee on Commerce
Rayburn House Office Building
Washington, D.C. 20515
(202) 225-2927
E-mail: Commerce@HR.House.gov

House Committee on Economic and Educational Opportunities
Rayburn House Office Building
Washington, D.C. 20515
(202) 225-2547

Chemical Industry Trade and Research Associations

The descriptions of the following associations come from their annual reports and brochures.

American Crop Protection Association (http://www.acpa.org)
1156 Fifteenth Street, N.W.
Suite 400
Washington, D.C. 20005
(202) 296-1585
The American Crop Protection Association represents the agricultural chemical industry.

American Industrial Health Council
2001 Pennsylvania Avenue, N.W.
Suite 760
Washington, D.C. 20006
(202) 833-2131

The American Industrial Health Council describes itself as promoting "the importance of sound science in regulatory decision-making on chronic human health hazards." Its members include Monsanto Company, the American Crop Protection Association, and the National Cotton Council.

Center for Emissions Control (http://www.sysnet.net)
2001 L Street, N.W.
Suite 506A
Washington, D.C. 20036
(202) 785-4374
E-mail: cec@sysnet.net
The Center for Emissions Control is a nonprofit, industry-sponsored organization that researches technologies on the use of chlorinated solvents, including perchloroethylene.

Chemical Industry Institute of Toxicology
Six Davis Drive
P.O. Box 12137
Research Triangle Park, North Carolina 27709
(919) 558-1200
The Chemical Industry Institute of Toxicology, founded in 1974, is a not-for-profit research laboratory which states as its mission to "provide an improved scientific basis for understanding and assessing the potential adverse effects of chemicals, pharmaceuticals, and consumer products on human health." It is financed primarily by approximately 50 corporations, including Monsanto.

Chemical Manufacturers Association
1300 Wilson Boulevard
Arlington, VA 22209
(703) 741-5000
The Chemical Manufacturers Association, the largest organization of U.S. and Canadian chemical companies, represents nearly 200 companies. In 1996, the CMA was engaged in a major lawsuit against the EPA, seeking to block the regulators' plan to increase the number of chemicals about which information must be reported to the public under the Emergency Planning and Community Right-to-Know Act.

Chemical Specialties Manufacturers Association
1913 I Street, N.W.
Washington, D.C. 20006
(202) 872-8110

The Chlorine Institute, Inc.
2001 L Street, N.W.
Suite 506
Washington, D.C. 20036
(202) 775-2790
According to its 1994 annual report, the Chlorine Institute "exists to support the chlor-alkali industry and serve the public." The Chlorine Institute, the report says, provides "management and support services to the Halogenated Solvents Industry Alliance and its product stewardship entity, the Center for Emissions Control." The Chlorine Institute's members include Dow Chemical, ICI, Occidental Chemical, PPG Industries, and Vulcan Materials.

Cosmetic, Toiletry, and Fragrance Association
1101 17th Street, N.W.
Suite 300
Washington, D.C. 20036
(202) 331-1770
Formaldehyde is commonly used as a preservative in cosmetics.

Halogenated Solvents Industry Alliance
2001 L Street, N.W.
Suite 506A
Washington, D.C. 20036
(202) 775-0232
The Halogenated Solvents Industry Alliance describes itself as an "affiliate of the Chlorine Institute."

Hardwood, Plywood and Veneer Association
P.O. Box 2789
Reston, Virginia 22090
(703) 435-2900

International Fabricare Institute
12251 Tech Road
Silver Spring, Maryland 20904
(301) 622-1900
The International Fabricare Institute represents the dry-cleaning industry.

Manufactured Housing Institute
1745 Jefferson Davis Highway
Suite 511
Arlington, Virginia 22202
(703) 558-0400

The Manufactured Housing Institute represents an industry that has been a particularly heavy user of formaldehyde products, including particleboard and paneling.

National Particleboard Association
18928 Premiere Court
Gaithersburg, Maryland 20879
(301) 670-0604

Neighborhood Cleaners Association (http://www.cns-nj.com/nca)
252 West 29th Street
Second Floor
New York, New York 10001
(212) 967-3002
The Neighborhood Cleaners Association is a worldwide membership organization of dry-cleaning establishments and associations.

Responsible Industry for a Sound Environment
1156 15th Street, N.W.
Suite 400
Washington, D.C. 20005
(202) 872-3860
RISE, an offshoot of the American Crop Protection Association, is the trade association of the specialty pesticide industry. Its membership consists of manufacturers, formulators, distributors, and other companies involved with specialty pesticides used in turf, ornamental, pest management, vegetation control, and other nonagricultural applications. In its publication *Pesticides in Your Environment*, RISE calls specialty pesticides "society's unsung heroes" and says that life would not be possible or enjoyable without them.

Synthetic Organic Chemical Manufacturers Association
1100 New York Avenue, N.W.
Suite 1090
Washington, D.C. 20005
(202) 414-4100
SOCMA is an association of companies that produce organic chemicals. Its several hundred members include du Pont, Hoechst Celanese, and other smaller makers of formaldehyde.

Industry-Supported Organizations

Council on Agricultural Science and Technology
4420 West Lincoln Way

Ames, Iowa 50014
(515) 292-2125
CAST was founded in 1972 as a not-for-profit organization whose mission is "to identify food and fiber, environmental, and other agricultural issues and to interpret related scientific research information for legislators, regulators, and the media for use in public policy decision making." Its members and underwriters include Ciba-Geigy Corporation, Monsanto Company, the American Crop Protection Association, the National Corn Growers Association, the National Cotton Council of America, and more than 200 other corporations and trade associations.

National Academy of Sciences
2101 Constitution Avenue, N.W.
Washington, D.C. 20418
(202) 334-2000
The National Academy of Sciences is a private, not-for-profit society of scholars that counsels the federal government. Among its many activities, the NAS researches agricultural, health, and environmental issues. Although most of its budget comes from the federal government, it also receives money from private sources, including Monsanto Company and the Chemical Manufacturers Association.

Society of Environmental Toxicology and Chemistry (http://www.setac.org)
1010 North 12 Avenue
Pensacola, Florida 32501
(904) 469-1500
E-mail: setac@setac.org
SETAC is a not-for-profit professional society founded in 1979, dedicated to interdisciplinary communication among environmental scientists and others interested in environmental issues, such as managers and engineers.

Organizations Working for Stricter Regulation of Alachlor, Atrazine, Formaldehyde, or Perchloroethylene

Consumers Union of the United States
101 Truman Avenue
Yonkers, NY 10703
(914) 378-2000

Environmental Defense Fund (http://www.edf.org)
1875 Connecticut Ave., N.W.
Suite 300
Washington, DC 20008
(202) 387-3500

Environmental Working Group
1718 Connecticut Avenue, N.W.
Suite 600
Washington, D.C. 20009
(202) 667-6982

Greenpeace, USA (http://www.greenpeace.org)
1436 U Street, N.W.
Washington, D.C. 20009
(202) 462-1177

Mothers and Others for a Livable Planet
(http://www.ecomail.com/ activist/mom.html)
40 West 20th Street
New York, New York 10011
(212) 242-0010

National Audubon Society (http://www.audubon.org/audubon)
666 Pennsylvania Avenue, S.E.
Suite 200
Washington, DC 20003
(202) 547-9009

National Coalition Against the Misuse of Pesticides
701 E Street, S.E.
Suite 200
Washington, D.C. 20003
(202) 543-5450

Natural Resources Defense Council (http://www.igc.apc.org/nrdc)
1350 New York Avenue, N.W.
Suite 300
Washington, D.C. 20005
(202) 783-7800

UNITE! (formerly the Amalgamated Clothing and Textile Workers Union
and the Ladies Garment Workers Union)
815 16th Street, N.W.
Suite 507
Washington, D.C. 20006
(202) 628-0214

Acknowledgments

The authors would like to thank the following individuals: Sheila Krumholz, Jacqueline Duobinis, and Dave Royce of the Center for Responsive Politics, and Anne McBride, Jane Metzinger, and Katherine Loos of Common Cause, for their assistance in assembling information on political contributions; Ken Niles, a librarian at the National Library of Medicine, for his help in retrieving scientific studies from a large volume of reference materials; and Andy Kivel of The Data Center in San Francisco, for research from difficult-to-find sources. At the Environmental Protection Agency, Patricia Sears, Robin Carnes, and librarians in the public docket rooms of the Office of Air and Radiation and the Office of Pollution Prevention and Toxics assisted with repeated requests for agency documents.

At *Newsday*, editors Anthony Marro, Howard Schneider, Charlotte Hall, Miriam Pawel, Robert Tiernan, and Ben Weller were supportive from the start and never lost their patience during what proved to be a long haul. At the *National Law Journal*, special thanks go to editor Ben Gerson and Washington bureau chief Marcia Coyle for offering the support, encouragement, and especially the time needed to complete this book.

Thanks also to Hillel Black, the editorial director of Carol Publishing Group, and Nina Graybill, our agent. Alison Frankel was also a valued adviser.

This book would not exist were it not for the generous support of four foundations: the Bauman Foundation, the Deer Creek Foundation, the W. Alton Jones Foundation, and the Giles and Elise Mead Foundation.

Finally, we thank our families and friends for keeping the faith as we went about our work.

Index